Sustainable Urban Design

Our future is in cities – it is thought that by the end of the twenty-first century three-quarters of the world's population will be urban. Making these cities healthy, vibrant and sustainable is an exceptional challenge, which *Sustainable Urban Design* addresses. The book sets out some of the basic principles of the design of our future cities and, through a series of carefully selected case studies drawn from UK and European experience by leading designers, illustrates how these ideas can be put into practice.

This work provides architects, engineers, students, clients, planners and public officials with a clear, coherent view of the important environmental issues in urban design. The focus is on the physical aspects of the urban environment: buildings (and their engineering systems, including the rapidly expanding field of solar electricity), landscaping, transport systems; and on the energy and material flows, including water and wastes. Particular emphasis is placed on exciting new energy systems, including the question of how we can move towards cities that are more enjoyable, more interesting and more green while using less energy, materials and water.

Randall Thomas Eur Ing PhD (Arch), CEng FCIBSE MASHRAE is a partner of Max Fordham LLP, Visiting Professor at Kingston University and lectures on sustainable urban design at the Architectural Association.

Randall Thomas came to sustainable design from a background of urban research in Paris and environmental research at Cambridge. Since 1981, when he joined Max Fordham LLP, he has been responsible for a variety of award-winning projects, including the Environmental Building at the Building Research Establishment. His current work includes major urban regeneration schemes in London and Belfast.

In addition to contributing numerous papers to conferences, he was co-author of *Design with Energy* (CUP, 1984) and editor of *Environmental Design* (Spon Press, 1999, second edition) and *Photovoltaics and Architecture* (Spon Press, 2001).

Sustainable Urban Design

An Environmental Approach

Edited by
Randall Thomas
Max Fordham LLP

Spon Press
Taylor & Francis Group

LONDON AND NEW YORK

First published 2003 by Spon Press
11 New Fetter Lane, London EC4P 4EE

Simultaneously published in the USA and Canada
by Spon Press
29 West 35th Street, New York, NY 10001

Spon Press is an imprint of the Taylor & Francis Group

Typeset in Futura Light by Wearset Ltd, Boldon, Tyne and Wear
Printed and bound in Great Britain by St Edmundsbury Press,
Bury St Edmunds, Suffolk

British Library Cataloguing in Publication Data
A catalogue record for this book is available from the British Library

Library of Congress Cataloging in Publication Data
Sustainable urban design / edited by Randall Thomas.
    p. cm.
  Includes bibliographical references and index.
   1. Urban ecology. 2. City planning—Environmental aspects. 3.
Sustainable development. I. Thomas, Randall.

HT241 .S8735 2002
307.76—dc21

2002026926

ISBN 0-415-28122-9 (hbk)
ISBN 0-415-28123-7 (pbk)

Contents

Contributors

Patrick Clarke (Msc, PhD) is a director of Llewelyn-Davies, where he leads much of the practice's planning policy and research on sustainability, design and urban quality. Much of his recent work has focused on exploring how urban areas can accommodate new development in ways that improve their quality and foster more sustainable patterns of urban living.

Tristan Couch joined Max Fordham LLP in 1997. He has an Honours degree in Civil Engineering from the University of Melbourne and a Masters degree by research in Passive Solar Housing Design from the Royal Melbourne Institute of Technology. He has worked on housing, concert hall and museum projects.

Bill Dunster, a graduate of Edinburgh University, worked with Michael Hopkins and Partners before forming Bill Dunster Architects. He has been involved with research projects working towards zero-energy urban buildings. These include Hope House, a prototype solar live/work unit, and, more recently, the urban village at Beddington in Surrey, from where he operates his practice.

William Filmer-Sankey is an associate at Alan Baxter & Associates, where he works in the urban design and conservation team.

Max Fordham (OBE MA (Cantab) FREng FCIBSE MConsE Hon FRIBA) is Principal of Max Fordham LLP, a practice of building services engineers that has contributed to some of the most advanced buildings of the last twenty years. He is Visiting Professor in Building and Design at the University of Bath and was President of the Chartered Institution of Building Services Engineers from May 2001 to May 2002.

Graham Haworth studied architecture at the Universities of Nottingham and Cambridge and founded Haworth Tompkins with Steve Tompkins in 1991. The creative philosophy of the practice centres on the potential of new architecture to bring about positive change and support a strong social and cultural agenda. Projects include the Royal Court Theatre and the Coin Street Housing.

David Levitt has thirty-five years' experience of designing housing in the UK, as a director of Levitt Bernstein Associates, and has been continually involved in research into housing policy and standards throughout that period. He is currently engaged in a number of major live regeneration schemes, which involve different forms of tenure diversification, refurbishment, redevelopment and resident consultation.

Mark Lewis studied at Sheffield University and qualified in 1993. Mark has been an architect at Levitt Bernstein since 1995, working on a mixture of theatre, museum and housing projects. At CASPAR, Mark helped develop many of the innovative prefabricated elements of the scheme, contributing a new approach to old design problems.

Richard Partington is Principal of Richard Partington Architects. His practice was established in 1998 and is working on regeneration and urban design projects in Belfast, Manchester and Kent.

Adam Ritchie is a building services engineer at Max Fordham LLP. He holds a BSc in Physics from York University and became a partner of Max Fordham LLP in 2002. His work includes the environmental and building services design for Cambridge University Botanic Garden Education and Interpretation Centre.

Alan Short trained at the Universities of Cambridge and Harvard, is Principal of Short & Associates, Professor of Architecture at the University of Cambridge and Fellow of Clare Hall, Cambridge.

Michael Sillén is a social scientist specialising in environmental policy. He has been working for the City of Malmö since 1995 and has a Bachelor's degree from the University of Lund, Sweden.

Randall Thomas (Eur Ing, PhD (Arch)) is a partner of Max Fordham LLP and has over twenty years of experience in an environmental approach to buildings and cities. He is Visiting Professor of Architectural Science at Kingston University and teaches sustainable urban design at the Architectural Association.

Robert Thorne is a senior associate at Alan Baxter & Associates, where he helps lead the urban design and conservation team. He is principal author of *Places, Streets and Movement* (DETR, 1998) and *Achieving Quality Streetscapes* (CABE/DETR, 2002). He holds a Visiting Professorship in Architecture at the University of Liverpool.

David Turrent trained at Manchester University and worked in private practice and local government before setting up ECD Architects in 1980. He has been responsible for numerous award-winning low-energy buildings and is a member of the RIBA Sustainable Futures Committee. He is also a member of the CABE Enabling Panel.

Christina von Borcke is an associate urban designer with Llewelyn-Davies. She holds a Bachelor's degree in Landscape Architecture from the University of British Columbia, Canada, and a Masters degree in Urban Design (Dist.) from Oxford Brookes University. Her professional work focuses on setting the scene for sustainable development through master planning and urban design.

Foreword

The Urban Task Force set out 'to establish a new vision for urban regeneration founded on the principles of design excellence, social well-being and environmental responsibility . . .'. *Sustainable Urban Design*, concentrating on environmental issues that are integral to the vision of the UTF, allows us to look beyond the individual building and appreciate the way in which urban form, transport, landscape, energy and buildings are all interrelated. Cities that are well designed are exciting, socially inclusive and economically thriving. Using solar energy, cities can produce energy rather than simply consume it.

It is encouraging to see planners, architects and engineers collaborating – together they can help reverse the process of fragmentation that has accelerated the decline of our cities. We need to regain an urban tradition while adapting it to embrace renewable energy sources.

The case studies in this publication are valuable signposts in that process. They mark achievements without shying away from the difficulties ahead. They emphasise the wide range of skills required to produce urban projects that enhance our quality of life.

Government, at many levels, has a key role to play in all of this. Its influence and initiatives should be wide-ranging – promoting the use of solar and wind power, facilitating public transport, supporting quality in the design of cities and funding creative education and research institutions.

The authors of *Sustainable Urban Design* form part of an increasingly persuasive group demonstrating that urban renaissance can make a key contribution to the environment. Their work takes us one step further to realising a twenty-first century characterised by environmentally responsible cities powered by solar energy. As more of us become city dwellers, we must ensure that our urban environment is as dynamic and as energy efficient as possible.

Richard Rogers
Richard Rogers Partnership

Preface

This book is a glimpse into a certain Utopia. It is divided into two major parts – the first sets out some introductory concepts and the second is a group of case studies. A number of appendices and a short glossary can be found at the end.

The figure below explains the structure of the first part.

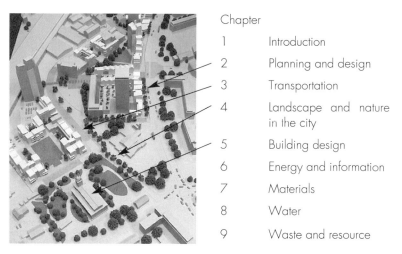

Chapter	
1	Introduction
2	Planning and design
3	Transportation
4	Landscape and nature in the city
5	Building design
6	Energy and information
7	Materials
8	Water
9	Waste and resource

The case studies draw on some of the best experience from the UK and beyond, such as the project in Malmö in Sweden.

We hope this work will stimulate all those interested in cities to develop their own Utopias – they are in short supply and in great need.

Good Luck!

Randall Thomas

Physics units, conversion factors and abbreviations

1. The unit of thermodynamic temperature in the SI system is the kelvin (K). For this reason derived units such as thermal conductivity are expressed as watts per metre kelvin (W/m K). However, the Celsius (°C) temperature scale is also in common use (the Celsius scale is also known as the centigrade scale). Absolute temperature in degrees kelvin is found by adding 273 to degrees Celsius. Thus, $30°C + 273 = 300 K$

2. Power
   W = watt (1 W = 0.86 kcal/h); kW = kilowatt; kWe = kilowatt of electrical output

3. Energy
   kWh = kilowatt hour; MWh = megawatt hour (1000 kWh); GJ = gigajoule (1 G) = 278 kWh)

4. Heat transfer coefficient
   $1 W/m^2 K$ (1 watt per m^2 of area per degree K of temperature difference)

5. Abbreviations
   dB = decibel; ha = hectare; Hz = hertz (1 Hz = 1 cycle per second); kcal = kilocalorie; nm = 1 nanometre

Acknowledgements

It is always a pleasure for an editor to acknowledge the contribution of so many to a book like this. If I may speak for all the contributors, the work relies on the support of our colleagues (and many other professionals) and clients who have helped create a climate of change that has enabled us to take the first steps towards a sustainable urban design.

Kingston University and the Architectural Association (AA) have been very helpful; the former, particularly in the person of Trevor Garnham, in arranging a conference on the subject in which a number of the chapters here were delivered as papers, and the latter, in the person of Michael Weinstock, for accepting my idea of a series of lectures on urban design. The students at both Kingston and the AA have assisted me enormously in the development of the text.

The book could not have been produced without the care, patience and skill of my colleagues – first and foremost, Seemi Gopinathan, closely followed by Cassius Taylor-Smith, Oak Taylor-Smith, Emma McMahon and Kitty Lux. Tony Leitch has elegantly transformed quite a few rough sketches into drawings. Caroline Mallinder and Michelle Green of SPM Press have been understanding, enthusiastic and friendly throughout.

I would like to thank Miriam Fitzpatrick, Trevor Garnham and David Lloyd-Jones, who kindly read the manuscript and then, with a number of well-penned pages, sent us back to our reference books and word processors to create more finely honed texts. Nonetheless, it goes without saying that any errors or lack of clarity should be attributed to the editor alone.

Nick Baker and Koen Steemers generously gave of their time to comment on a number of technical points.

My inspiration for this book comes from Paul and Percival Goodman's *Communitas* and Lewis Mumford's *The City in History*. No editor could have had better guides.

Paul Klee's drawings illustrating "natural and artificial measurement" one day led to my city. And I need to apologise to another artist, Rembrandt, for an elephant.

Finally, and symbolically, I would like to thank Dave Bodenham, who let us use his delightful photograph of a sheep. Profits from this photograph go to a charity for survivors of Chernobyl with which he is involved. Anyone wishing to make a donation can contact him at bodenham@hotmail.com.

Note to readers

One intention of this publication is to provide an overview for those involved, professionally, as students, or in any other way, with cities. It is not intended to be exhaustive or definitive and it will be necessary for users of the information to exercise their own professional judgement when deciding whether or not to abide by it.

It cannot be guaranteed that any of the material in the book is appropriate to a particular use. Readers are advised to consult all current Building Regulations, British Standards or other applicable guidelines, Health and Safety codes and so forth, as well as up-to-date information on all materials and products.

PART ONE

1

Introduction

Randall Thomas

1.1   Sustainability

Sustainability is about poetry, optimism and delight. Energy, CO_2, water and wastes are secondary. The unquantifiable is at least as important as the quantifiable; Louis Kahn said "the measurable is only a servant of the unmeasurable" (1) and ideally the two would be developed together – in Jonathan Swift's terms, this is a modest proposal.

The goal of this book is to identify the major issues in making cities environmentally sustainable by, say, 2020. Predictions vary but it seems likely that over two-thirds of the world's population will be living in urban areas towards the middle part of this century. Cities offer tremendous opportunities for contact, employment, excitement and interest, which attract many of us. They can also create problems of over-intensive resource use, of congestion, of noise and pollution but these can be addressed, in part, through design. Urban success depends on recognising trade-offs and getting the balance right.

It is vital that we evolve towards sustainability in urban form, transport, landscape, buildings, energy supply, and all of the other aspects of vibrant city living. Part of this will involve making cities more suitable for people, and so shifting away from the previous policy of cities being for cars. Creating environments with pedestrians, cyclists and public transport in mind is a key aspect of sustainable development.

One model for this transition is the ecosystem, with its robust and stable systems. However, in addition to an ecosystem's variety, diversity, redundancy and richness, it would be nice to have poetry, whimsy, playfulness and excitement. Aristotle's view was that a city should be designed to make its people secure and happy (2).

This book is modestly meant to inspire rather than to be prescriptive (indeed given so much that is not predictable, it could hardly be otherwise). Ideas of planning, space and form are a light backdrop to much of the discussion but our built environment suffers enough at present from those who were too sure of their solutions and those who would overplan and, thus, overconstrain development. The contributors, however, do believe that an integrated approach is needed. Density and the means of moving about the city are related. Landscape affects buildings. Noise influences the ventilation system selected and thus energy use. In turn the energy use currently results in increased atmospheric pollution at power stations, which affects our health. Similarly, the built form affects access to sunlight and this influences both the energy use and our well-being.

Some general themes run through the book. One is that appropriate solutions often depend on an understanding of the context; environmental, historical, social and so on. A second is that the appropriate scale for solutions is something larger than the individual building – it could be the block, the neighbourhood, the city, the region. Another is that solutions that require fewer resources rather than more are more likely to be robust. And so for energy, the first step is to reduce demand and then to examine how to meet it. In terms of movement in cities and towns, the robust solution is the dense walkable community, which does not have a very high demand for either public or private transport. There is also a view that passive solutions will be best. Things that move in the urban world tend to be less robust and require more maintenance. This is true whether one looks at cars, or London Underground's escalators, or pumps for heating systems.

Our subject has essentially been cities in temperate climates that are neither bitterly cold in winter nor desperately hot and humid in summer. Our emphasis has been on new build, in part, because that is where most of the contributors have had the greatest opportunities to develop new approaches. Improving the existing building stock is an equal challenge and, fortunately, a number of the ideas discussed below are applicable to upgrading existing buildings. Our case studies draw on some of the best experience from the UK and beyond, such as the case of Malmö. Many of our examples are from brownfield sites where new buildings have often led to the regeneration of neighbourhoods. The increased densities that are likely to help us reduce CO_2 emissions will depend in part on our ability to recycle creatively both land and buildings.

Reference is often made to the three very interdependent aspects of sustainability: social (and although it is clear that sustainability is dependent on communities for its success, developing the social dimension is often an arduous process), economic and environmental; this book will concentrate on the environmental. The emphasis on this aspect arises naturally – it is the field that most of the contributors are working in. A (readable) book is, of course, also necessarily limited in length and one needs to choose one's focus. But, on a light note, it could be said that environment is the easiest aspect (in spite of the resistance to environmental improvement by some companies and governments, e.g. the current US administration under President George W. Bush) because it is much simpler to assess, for example, whether CO_2 emissions have been reduced by a scheme than whether a scheme will successfully lead to economic regeneration.

There is also a fourth aspect of sustainability, which is individuality – it creates space for the unexpected and the extraordinary as well as centring itself, in the best humanist tradition, on the person. If the social dominates the individual, all suffer from the deadening mediocrity that one finds outside the historical centres of cities as diverse as St Petersburg, Paris and London. This theme of uniqueness surfaces in urban design, landscaping and in individual buildings and will be found throughout the book. The exceptional towers of De Montfort University's Queens Building (Figure 1.1) (see Chapter 15) are an example. The structures are both symbolic of urban regeneration and functional, serving as an integral part of the ventilation system.

Sustainable cities will have a rich set of interconnections or they will not be sustainable. For example, a city for walkers and cyclists needs more visual variety, more diversity, more "accidentals" (in the musical sense) in its street patterns and its buildings because the pace is slower and the mind both desires more and can take in more. We need to develop a rhythm in the city that will include places we can enjoy; this rhythm will be about moving – and stopping. This will help us return to cities designed for people, rather than for cars.

A city's "accidents" include such extraordinary views as this one down a narrow mews in London (Figure 1.2), which combines dramatic changes of scale, buildings from different centuries, backs and fronts. One small image indicates the diversity and vitality of real cities. "Accidents" are part of the normal pattern of our historical cities and appear to be anathema to many modern planners obsessed by regular patterns, they are part of our past,

Figure 1.1
De Montfort University, Queens Building

our memory and, so, part of our poetry. They tell us that all was not designed in an instant and so reinforce our sense of time. A delightful exception to conventional planning is the irregular streets of a new development in Malmö (Figures 1.3 and 17.1).

In many urban areas, public space, including parks and streets, is more than half the total area of land (3). The buildings provide us with homes and workplaces and with commerce, industry and leisure. The space in between the buildings (see Figure 1.4) provides vitality, light, amenity, room to travel and room to rest. Landscape is essential – plants soothe us and improve the microclimate. Our open space also allows for wildlife, thus promoting biodiversity. It may incorporate vegetable gardens and reed beds for waste treatment.

The interrelationship of three of the key factors in environmental sustainability can be viewed simplistically as a triangle (Figure 1.4c). One apex is form/density, a second is movement/transport and the third is buildings/ energy (use and production). We are only beginning to start to think about how these factors (and others such as landscape and social conceptions of privacy) can work together. This approach is in its infancy and the book describes some of the first steps being taken.

Urban form will affect energy use for transport and buildings. There is some agreement that the energy per capita for transportation decreases (Figure 1.4d) (4), but what happens to the energy required for buildings (Figure 1.4e) and the energy that the buildings can produce (Figure 1.4f)? These are important issues for environmental sustainability – and ones to which we return.

Figure 1.2
View down a narrow street in central London

1.2 The background

There is considerable scientific evidence that the earth's temperature is rising and will continue to do so as a result of human activities. These activities include particularly the burning of fossil fuels – coal, oil and gas – for our buildings and our transport, manufacturing and agricultural systems and they result in increased CO_2, which contributes to global warming. Climate change affects us all – we should act to reduce its impact by changing from fossil-fuel-based systems to renewable-energy-based ones.

The likely negative impacts of global warming (which will probably outweigh the positive effects) include increased storms, flooding, droughts and the probable destruction of some ecosystems. In urban areas, there is a "heat island" effect resulting from the production and accumulation of heat in the urban mass (5). Cities can be several degrees warmer than their surroundings. Figure 1.5 shows that on a night with a clear sky in May 1959 the temperature across the London region over a diameter of about 40 km varied from 4.4°C to 11.1°C, a heat-island effect of 6.7°C (6).

The heat-island effect will lead to temperature rises being more marked; air pollution in cities may increase; drainage systems may need to be altered to cope with periods of higher rainfall (7). Global warming will probably lead to social, political and economic disruption. It seems wise to abide by what is known as the "precautionary principle", which maintains that we should take action now to avoid possible serious environmental damage even if the scientific evidence for action is inconclusive (8), and design our cities to reduce their CO_2 production significantly. (It should be noted that for many there is no question of the scientific evidence for global warming as a result of human activity being inconclusive.)

The technology to reduce climate change is largely available. We need more resolve, more government support and more individual initiative.

New Buildings:
- Housing
- Building for other purposes
- Building under construction
- Planned building

Figure 1.3
Malmö street pattern

Figure 1.4
Urban space

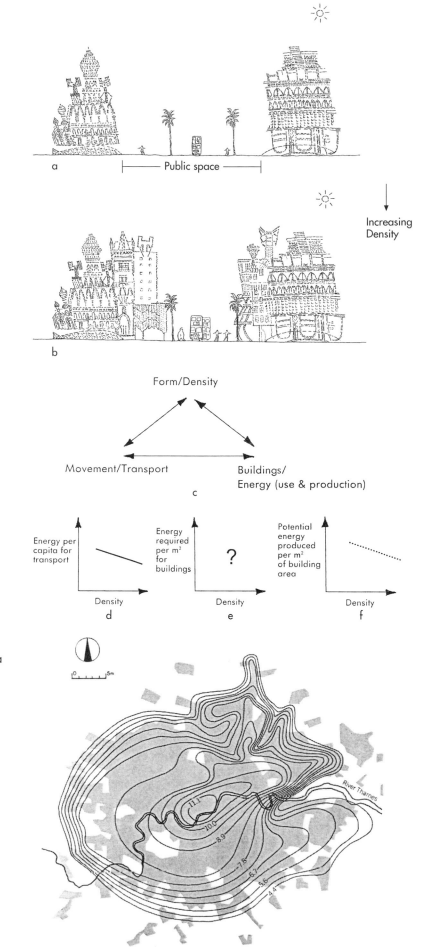

Public space

Increasing
Density

Form/Density

Movement/Transport

Buildings/
Energy (use & production)

c

Energy per
capita for
transport

Density

d

Energy
required
per m²
for
buildings

?

Density

e

Potential
energy
produced
per m²
of building
area

Density

f

Figure 1.5
Urban heat island effect in London on a
clear night (figures are in °C)

River Thames

11.1
10.0
8.9
7.8
6.7
5.6
4.4

Approximately 25 per cent of the UK's final energy demand is used in industry (including agriculture), 28 per cent in the domestic sector, 34 per cent in transport and 13 per cent in the service sector (9). Of course, in all these sectors much of the energy use is in buildings – in the European Union, buildings account for about 40 per cent of consumption (10). Table 1.1 shows the current source of energy supplies in the UK.

There is an urgent need to increase the contribution from renewables, and governments are (slowly) beginning to respond; the UK, for example, has a (modest) target of renewable energy providing 10 per cent of electricity supplies by 2010. How the renewable contribution could be increased in the urban context is discussed below, particularly in Chapter 6.

In beginning to consider cities it may be useful to think of systems. When looking at energy flows in natural communities ecologists do this regularly. A system might be defined as "regularly interacting and interdependent components forming a unified whole (12)." Figure 1.6 shows how the new housing being developed at Coopers Road in Southwark, London (see Chapter 11), might be viewed as a small urban system and gives very approximate energy and material flows. A very simple, but useful, approach to sustainability is to look at such flows and ask: where does it come from, how does it get there, who looks after it, what does it do, where does it go? (13). Of course, it will be more productive to work with a somewhat larger system since it is clear that, in energy and environmental terms, urban areas operate within the context of the city as a whole, the city works with the region and the regions function at a national or international level. None the less, this gives us a way of thinking about some of the issues. There is also the beginning of a lively debate about whether cities should be able to export their problems, for example, waste disposal and energy supply to the region. The electricity supply for the citizens of New York comes in part from a nuclear power station just north of the city that disfigures an area of exceptional natural beauty along the Hudson River. For how long can this go on?

Very broadly, 25 per cent of the primary energy is for the buildings, 40 per cent for transport and 35 per cent for food (see also Chapter 16). The resources, which are discussed in more detail in Chapter 6 and Appendix A, are a set of opportunities. Thus, the incident solar radiation is 16,000,000 kWh/y. Of course, our systems won't be 100 per cent efficient. But if we assumed that they were 10 per cent efficient, we would have 1,600,000 kWh/y from solar energy or roughly 1.5 times our primary energy demand for heating very energy-efficient housing. This is encouraging but a close look at the figures suggests a high level of sustainability in cities will not come easily. For example, CO_2 production is high and uptake low; rainfall is low and use is high. On a worldwide level, solar energy incident on the earth's surface is about 15,000 times the demand for energy (22).

Table 1.1
UK energy supplies (11)

Item	Approximate percentage of supply (a)
Oil	38.5
Coal	15
Gas	38
Nuclear	7
Renewables (solar, wind, etc.)	1.5

Note
a. Total energy supply is approximately 10,700 million GJ/y (1 GJ = 278 kWh).

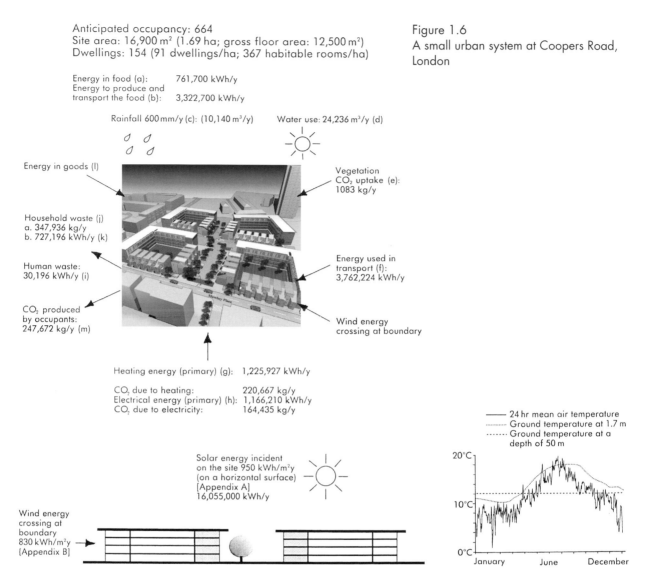

Anticipated occupancy: 664
Site area: 16,900 m² (1.69 ha; gross floor area: 12,500 m²)
Dwellings: 154 (91 dwellings/ha; 367 habitable rooms/ha)

Figure 1.6
A small urban system at Coopers Road,
London

Energy in food (a): 761,700 kWh/y
Energy to produce and
transport the food (b): 3,322,700 kWh/y

Rainfall 600 mm/y (c): (10,140 m³/y)   Water use: 24,236 m³/y (d)

Energy in goods (l)

Household waste (j)
a. 347,936 kg/y
b. 727,196 kWh/y (k)

Human waste:
30,196 kWh/y (i)

CO₂ produced
by occupants:
247,672 kg/y (m)

Vegetation
CO₂ uptake (e):
1083 kg/y

Energy used in
transport (f):
3,762,224 kWh/y

Wind energy
crossing at boundary

Heating energy (primary) (g): 1,225,927 kWh/y

CO₂ due to heating: 220,667 kg/y
Electrical energy (primary) (h): 1,166,210 kWh/y
CO₂ due to electricity: 164,435 kg/y

Solar energy incident
on the site 950 kWh/m²y
(on a horizontal surface)
[Appendix A]
16,055,000 kWh/y

Wind energy
crossing at
boundary
830 kWh/m²y
[Appendix B]

———— 24 hr mean air temperature
············ Ground temperature at 1.7 m
········ Ground temperature at a
depth of 50 m

20°C

10°C

0°C

January   June   December

Notes

a. This is an approximate figure for the energy content of the food based on (14).
b. (15).
c. The rainfall figure is an approximate average.
d. With water-saving measures (see Chapter 8) it should be possible to reduce household water consumption to 36.5 m³/person/year or 24,236 m³/y for the site.
e. A very rough estimate is that the grass and small trees of Coopers Road might assimilate 50 per cent of the CO₂ of cultivated fields and forests thus giving a figure of about 0.55 kg CO₂/m²y of planted area. The uptake is based on planted areas of 1490 m² and lawn areas of 479 m² or a total of 1969 m², roughly 11 per cent of the site area (16).
f. An average person in the UK has a transport energy consumption of about 5666 kWh/y; approximately 82 per cent of this is for car travel, 5 per cent for rail and 7 per cent for bus (17). This average figure may very well be significantly lower in urban situations such as Coopers Road. For an analysis of an inner-city area in Bristol see the Rickaby reference under Further Reading.
g. Gas delivered energy: 1,103,334 kWh/y; conversion factor: 0.90 (18); therefore primary 1,225,927.
h. Electricity-delivered energy: 349,863 kWh/y; conversion factor: 0.30 (19) (note that more precise calculations will need to check if this factor has altered significantly); therefore primary 1,166,210. (Note that delivered energy is assumed to be equal to useful energy.)
i. This is calorific value and is estimated as 4 per cent of the calorific value of the food (20).
j. Household waste: average person 524 kg/y (see Chapter 9).
k. Waste energy content (21).
l. Not quantified here. This term will include the energy in goods (papers, tins, packaging, etc.). A thorough analysis will need to take into account such subtleties as the non-consumed part of food, e.g. vegetable parings, bones, etc.
m. Estimated from an energy intake of 10,000 kcal per day per person; 373 kg CO₂/person/y.

One can look at the demands as the environmental "footprint" of the site and the challenge is to reduce it over time and increase our "footprint" of the use of ambient energy sources concurrently as shown in Figure 1.7.

It will also be helpful to develop a way of thinking about how the city will change with time and to set out a clear and coherent strategy for improvement. We, of course, need to agree on a time-scale for our activities.

Optimists are aiming for environmentally friendly cities in 2020. Realists sometimes cite 2050. We don't even want to tell you what the pessimists say. Throughout the book there is pragmatic advice on how to reduce our impact and exploit our opportunities.

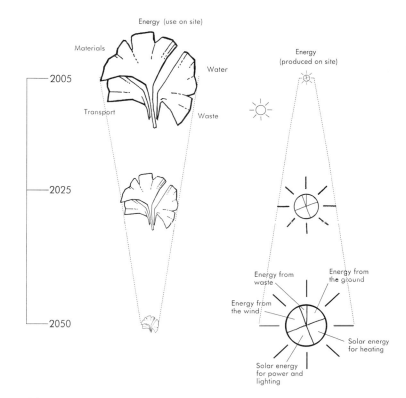

Figure 1.7
Evolving environmental "footprints"

Table 1.2
A strategy for looking at how cities might evolve – key indicators and sliding scales

	Light Green (2000)	Dark Green (say, 2020)
1. Reduction in space and hot water heating demand	0 → 50—o— 100	0 50 —o— 100
2. Reduction in electrical-energy demand	0 50 → —o⊢ 100	0 50 —o⊢ 100
3. Provision of hot-water demand using solar energy	0 → 50—o 100	0 50 —o⊢ 100
4. Provision of electrical-energy demand using photovoltaics	0 → 50—o 100	0 50 —o⊢ 100
5. Recycling of rain water	0 → 50—o⊢ 100	0 50 —o⊢ 100

a

b

Figure 1.8
Solar shading
a. Horse and shadow
b. Devil's Passage, Paris

Integral to this is an appreciation of how to "look" at an urban site. This is best viewed as a complete analysis of assets, including the social aspects of the neighbourhood (its meeting places, communication routes, safe areas), circulation routes (both for pedestrians and for vehicles), historical monuments, important places, good architecture, interesting geology, trees and so forth. The principal environmental concerns are touched on below.

1.3 Site analysis

Solar energy is our starting point and there is a long history of philosophers and architects attempting to find, or, indeed, impose, an "optimum" form that would provide for residents a healthy environment with daylight, air and space. In the twentieth century Le Corbusier, Frank Lloyd Wright and Lubetkin (see Figure 6.16) were three of many influenced by such concerns. It seems reasonable that the form of a city grows from its environmental and social, in the broadest sense, requirements. A key element in this is maximising the solar potential of the site. To do this, a knowledge of solar geometry will be of help.

Figure 1.8a shows the shadow cast by Barry Flanagan's delightful horse in a Cambridge College garden; the solar elevation angle at noon on 22 September is about 38°. Figure 1.8b shows the intriguingly named Devil's Passage in Paris. Buildings in London of similar height (about 14 m) and separation (about 9 m), orientated with the long axis E–W, will have most of the south façade in sun at noon in September. (Of course, earlier and later during the day in September and later in the year the façade will be more shaded.)

Urban designers need to take into account the sun's path for many reasons, ranging from the poetic effects of sunlight, to the bringing in of daylight to ground-floor living rooms, to ensuring that roof-mounted photovoltaic panels are not overshadowed.

Fortunately, tremendous progress is being made in modelling the urban environment. Figure 1.9 shows an analysis of yearly solar irradiation on part of the San Francisco financial district (23). Unfortunately, black-and-white reproduction doesn't do the original justice and the reader is referred to the website for a number of striking images.

A more detailed discussion of solar geometry is found in Appendix A. Solar considerations in planning and design are discussed particularly in

Figure 1.9
Yearly solar irradiation on surfaces in the San Francisco financial district

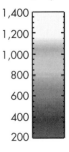

Irradiation scale
kWh/m²y

1,400
1,200
1,000
800
600
400
200

Chapters 5 and 6. The other principal items to consider are wind, air quality, temperature, rain and biodiversity. Water can also be very important (see Chapter 6).

The wind crossing a site is potentially beneficial. It can assist in the natural ventilation of buildings (see Chapter 5), remove pollutants and heat, and be a potential source of energy (Chapter 6). Urban environments may also exacerbate the adverse effects of the wind. (Guidance on designing with the wind is given in reference 24). One of the reasons for the irregular street pattern in Malmö (Figure 1.3) was to reduce the effect of the sea wind off the Oresund. Wind-tunnel testing, as shown in Figure 15.5 where the Contact Theatre in Manchester and its immediate urban setting was being studied, allows us to create a more amenable environment both for buildings and for people (as pedestrians and idlers).

Air quality affects our comfort, our health and our ability to use natural ventilation (see Chapter 5). As the inhabitants of a number of Californian and other cities throughout the world know, high levels of pollution can also shield the sun and reduce daylight levels.

With regard to temperature, as we have seen, the urban climate is warmer. Microclimates exist throughout the city and through design there is enormous potential for creating sunnier, warmer spaces in winter and cooler, shaded ones in summer. The use of temperature differences between ground and air (Figure 1.6) is also a potential source of low-CO_2 cooling and heating. Chapter 6 and Appendix A provide more detail.

The rain falling in an area is a precious resource but, as parts of England have seen recently, if it is not managed well it carries a risk of flooding. Chapters 4, 8 and 9 provide more information.

Sound (and noise) is almost a defining feature of urban life. Chapter 5 and Appendix D discuss this further.

Of course, the impact on the environment is not simply one of energy and material flows. There may be a reduction of biodiversity. This is because the number of species tends to increase with the availability of area, thus a loss of space through building is likely to be deleterious. We need to mitigate these effects. Note the difference with energy where it is quite conceivable that through PVs we may produce more energy than the building uses, particularly in the future when technological developments reduce demand and increase supply efficiencies. Maintaining (or, we hope, increasing) biodiversity is considered in more detail in Chapters 4 and 17.

1.4 Conclusion

Sustainable urban design is vital for this century – it is not too much to say that our health, welfare and future depend on it. Achieving a transition to a sustainable, solar society will involve us all – all should share in the environmental, social and economic benefits.

We will need to develop flexible ways to "shape" and design our future cities. Land-use planning can contribute to reducing travel demand and, thus, CO_2 emissions and, similarly, it can increase our use of the solar potential of the site.

Sustainable urban design is, in part, about balance. There are many good reasons for cities that are dense, have mixed uses and are varied. But such cities will need to manage potential conflicts between varying conceptions of urban form and living, between public transport and individual cars, between public distribution of energy supplies and private control, between man-made environments and more natural ones, and many others. This challenge is about our futures – it is demanding and exciting. Its outcome will not be one solution but many. There will be some clarity but, more likely, if we are successful, our cities will be "magnificently equivocal" (25).

REFERENCES AND NOTES

1. Kahn, L. (1969), "Silence and light: Louis I. Khan at ETH (Zurich)", in Heinz Ronner, Sharad Jhaveri, Alessandro Vassella, Louis I. Khan: complete work 1935–74, Institute for the history and theory of architecture. The Swiss Federal Institute of Technology, Zurich, 1977, pp. 447–449.

2. Cited in Collins, G. and Collins, C. (1986), Camillo Sitte: The Birth of Modern City Planning. Rizzoli, New York.

3. The Urban Task Force (1999), Towards an Urban Renaissance. E&FN Spon, London, p. 56.

4. Fouchier, R. (1997), Les densites urbaines et le developpement durable. Editions du SGVN – Secretariat General du Groupe Central des Villes Nouvelles.

5. In cities, heat is produced by people and equipment and this heat and solar radiation are absorbed in the buildings and streets; there is less heat loss by evaporation and wind speeds are generally lower so heat is less easily carried away.

6. Dodd, J. (1989), Greenscape 5 Green Cities. Architects Journal, 10 May, pp. 61–69.

7. Anon. (2001), Engineering and Physical Sciences Research Council and UK Climate Impacts Programme. The Impacts of Climate Change in the Built Environment, Transport and Utilities.

8. For an alternative view see Ridley, M. (2001), Technology and the environment: the case for optimism. RSA Journal, 2/4, pp. 46–49.

9. Anon. (2000), Energy Projections for the UK; Energy Paper 68. DTI, London.

10. Santamouris, M. (2001), Urban reality. Renewable Energy World, 4(6), pp. 87–95.

11. Abstracted from DTI Digest of UK Energy Statistics (2001), DTI, London.

12. Cited in Odum, E. (1971), Fundamentals of Ecology. Saunders, London.

13. Mackle, D. (2001), The Children's Ecocity. Town and Country Planning, December, p. 326.

14. Mellanby, K. (1975), Can Britain Feed Itself? Merlin, London.

15. Pearce, F. (1997), White Bread is Green. New Scientist, 6 December, p. 10. Based on 5004 kWh/person/year (18,000 MJ person/year).

16. See Bolin, B. (1970), The energy cycle of the biosphere. The Biosphere, Scientific American, W. H. Freeman, San Francisco.

17. Banister, D. (2000), The Tip of the Iceberg: Leisure and Air Travel. Built Environment 26 (3) pp. 226–235.

18.   Thomas, R. (1999), Environmental Design. Spon, London, p. 84.

19.   Ibid.

20.   Durnin, J. V. G. A. and Passmore, R. (1967), Energy, Work And Leisure. Heinemann, London.

21.   Houghton, J. (1993), Royal Commission on Environmental Pollution, Seventeenth Report: Incineration of Waste. HMSO, London.

22.   Hoagland, W. (1995), Solar Energy. Scientific American, Key Technologies for the 21st Century. September, pp. 136–139.

23.   Mardaljevic, J. (2001), De Montfort University. ICUE: Irradiation mapping for complex urban environments. Website: www.iesd.dmu.ac.uk/~jm/icue/

24.   Anon. (1990), Climate and site development. Part 3: Improving microclimate through design. BRE Digest 350, BRE, Garson.

25.   The term "magnificently equivocal" was applied to *American Pastoral* by J. Savigneau in "Une Tragedie Ordinaire", Le Monde, 23 April 1999, p. vii.

FURTHER READING

Anon. (2000), Urban Design Compendium. Llewelyn Davies for English Partnerships and the Housing Corporation, London.

Anon. (2001), IPCC 3rd Assessment Report. Intergovernmental Panel on Climate Change, Geneva.

Behling, S. and Behling, S. (2000), Solar Power. Prestel, Munich.

Department of Transport, Local Government and the Regions/Commission for Architecture and the Built Environment (2001), Better Places to Live: by design. Thomas Telford Publishing, London.

Fitch, J. M. (1975), American Building 2: The Environmental Forces That Shape It. New York, Schocken.

Girardet, H. (1999), Creating Sustainable Cities. Green, Dartington.

Rickaby, P. (2002), in Buildings. Architecture Today, April, p. 54.

Rogers, R. (1997), Cities for a Small Planet. Faber, London.

Smith, P. (in preparation), Sustainability at the Cutting Edge. An account of the 2001 RIBA conference, Architectural Press, London.

The Sustainable Urban Neighbourhood. 41 Old Birley Street, Hulme, Manchester, M15 5RF.

WEBSITE

International Energy Agency: www.iea.org

2

Urban planning and design

Patrick Clarke

2.1 Introduction

This chapter seeks to draw out some of the planning and design principles that underpin attractive, successful and sustainable urban environments. The aim is to encourage a wider understanding of how basic layout and design principles can help create a robust urban form around which innovative design and the application of emerging construction and service technologies can flourish.

A key starting point is the recognition that, in large measure, the way in which people and goods move around urban areas determines their structure and how they function. When the means of movement changes, so too do patterns of human activity with direct consequences for the planning and design principles that guide the development and renewal of urban areas.

Most recently, the car has turned the established structure of urban areas inside out by drawing jobs, leisure and shopping away from traditional centres to locations that are easily accessed by car. Roads have been designed according to vehicle flows and speeds and buildings pushed back from the pavement to make space for parking, and, perhaps for the first time in history, children have been unable to play safely in the streets around their homes.

Now, in seeking to reduce our reliance on the car and to re-establish walking, cycling and public transport as the preferred means of moving around urban areas, we are again engaged in a process of change. This has profound implications for planning and design thinking at all levels; from how we think about the structure of towns and cities and where we locate new development right through to how we lay out individual street blocks and design the buildings within them.

Figure 2.1
A historic approach to urban transport in a contemporary form: Le Tram, Strasbourg

Central to much of the discussion in this chapter is the belief that while designing for the car took us into uncharted territory (often with disastrous consequences) there are clear historical precedents that can guide us in seeking to create places that work around pedestrian-, cycle- and public-transport-based movement. Indeed, it is ironic that while much energy is expended debating approaches to higher density, mixed-use environments we are almost unconsciously surrounded by places that exhibit many of these same attributes and that have succeeded in meeting the needs and aspirations of successive generations.

This introduces a second main thrust of this chapter: that people's requirements and technology change faster than places. Bespoke places or buildings that are designed to fulfil a specific function run the risk of becoming obsolete as requirements or technology change. A key point in creating places is to remember that while a street may have a life of 1,000 years or more, and a building perhaps 200 years, building services may have a life of just 25 years. It is crucial that these can be upgraded during the buildings' lives as technology develops.

Figure 2.2
A robust and enduring Victorian suburb: Jesmond, Newcastle

Many Victorian neighbourhoods, such as that illustrated in Figure 2.2, have seen the coming and going of different approaches to space heating, responded to changing household size through subdivision, adapted to mass car ownership better than many areas designed for the car and today are ready to welcome back the bus or tram with many of the essential attributes of density and street layout.

2.2 Sustainable urban structure

Thinking about sustainable urban structure begins with the urban region; the town or city and its rural and/or coastal hinterland. The sustainability of each is interdependent. The town or city depends on its hinterland for food and water, clean air and open space and, looking to the future, biomass for fuel. The hinterland depends on the town or city as a market for its produce and for employment and services but is also affected by urban waste and pollution. Sustainable planning demands a more holistic and integrated approach to the urban region, which recognises the interdependence and potential of both town and country.

At the level of the town or city the walkable community or urban village provides a fundamental building-block in creating a sustainable urban form. The concept is of a polycentric urban structure in which a town or city comprises a network of distinct but overlapping communities, each focused (depending on the scale of the urban area) on a town, district or local centre, and within which people can access on foot most of the facilities and services needed for day-to-day living. (1) Each of these communities is defined by the walking catchment or "ped-shed" around the centre. This is generally taken to be c. 800 m, equating to a 10-minute walk.

Figure 2.3 illustrates this concept in relation to a large metropolitan area in which the polycentric structure developed typically as the city spread outward, engulfing surrounding villages and towns and with new centres being created along new railway corridors. It shows what could be described as "a centres and routes" model, which can be observed in London, Birmingham or Manchester, for example. In this model, town centres are the principal community focus but there are also linear communities developing along the main movement routes between centres and especially along the principal routes to the city centre. In other places different structures can be seen reflecting differences of geography, landform and economy. In Belfast, for example, the structure is of linear communities that have developed along the principal arterial routes leading from surrounding towns to the city centre. This structure is described further and illustrated in Chapter 12 (see Figure 12.2).

In neither case is all of the urban area within the walking catchment of a centre. In general, the proportion of remote areas (that is, those lying beyond walking distance of a centre) increases with distance from the city centre, reflecting both diminishing densities of population and more widely spread movement routes.

2.3 The walkable community

Figure 2.4 (2) looks in more detail at the characteristics of a typical neighbourhood. A number of planning and urban-design principles can be drawn out:

- Shops and services tend to be focused along a main street running through the heart of the neighbourhood, at the convergence of movement routes and around key facilities such as a railway station. The degree to which shops and services spread outwards into

Figure 2.3
A polycentric urban structure of walkable communities

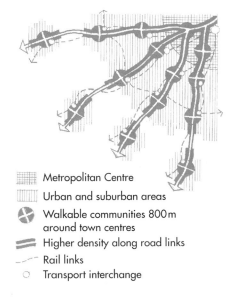

 Metropolitan Centre
Urban and suburban areas
Walkable communities 800 m around town centres
Higher density along road links
Rail links
Transport interchange

Figure 2.4
Attributes of a walkable community

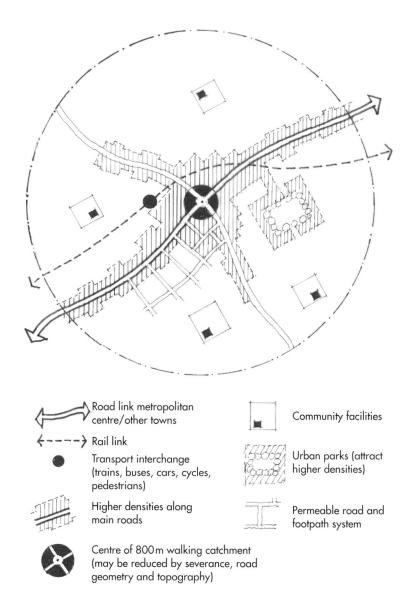

Road link metropolitan centre/other towns		Community facilities	

Road link metropolitan centre/other towns

Rail link

Transport interchange (trains, buses, cars, cycles, pedestrians)

Higher densities along main roads

Centre of 800 m walking catchment (may be reduced by severance, road geometry and topography)

Community facilities

Urban parks (attract higher densities)

Permeable road and footpath system

surrounding streets is a function of the scale and role of a centre, the density of population (and spending power) within its catchment and the degree of competition from neighbouring centres.

- Community facilities such as schools, health centres and open spaces are distributed around the neighbourhood reflecting more localised catchments and their greater requirements for space.
- The neighbourhood provides a wide range of different housing opportunities not just in terms of dwelling size but also in terms of affordability and tenure. This provides the basis for a mixed community representative of society at large rather than having a narrow social focus.
- Housing densities are highest around the edges of the town or district centre, along the principal transport routes leading to neighbouring centres and overlooking parks, waterfront areas and other amenities. Densities reduce towards the edge of the walking catchment.
- Movement routes are shared by cars, buses (or trams), cyclists and pedestrians and go through the centre rather than around it as well as through residential neighbourhoods.

Figure 2.4 also highlights a number of issues concerning existing neighbourhoods and their ability to accommodate new development and change:

- The potential for new infill development is often greatest in the "shatter zone" or "interface area" between the established retail, commercial and administrative centre and the consolidated residential hinterland. Here, land is often in marginal or short-term use, reflecting uncertainty or speculation over the optimum future use, outdated development-plan zonings or uncertainty over proposals such as for new roads. Such areas therefore often offer rich potential for new housing and mixed-use development.
- The opportunities for new mixed-use development are likely to be greatest within and on the fringes of the town/district centre and along the main movement routes to other centres. These areas often have a mixed-use character and can offer the greatest potential to support a range of uses, including shops, offices, leisure and housing.

2.4   Planning and design implications and opportunities

This idealised concept of walkable communities provides the basis for planning and design thinking at a variety of levels and in particular about how new-development and urban-management approaches can reinforce and strengthen a sustainable urban structure. This includes:

- directing new shopping, leisure and commercial development towards existing centres and ensuring that off-centre development does not undermine the viability of existing town and district centres;
- enhancing the attractiveness of existing town and district centres by extending the range of facilities and services they offer and improving the quality of their physical environment. Figure 2.5 (3) illustrates the potential to improve the quality of an important district

Figure 2.5
Regeneration of an existing district centre

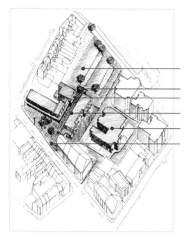

Public car park

Public conveniences
Library
Pedestrianised shopping parade
Derelict workshops/warehouses
Ill-defined and poorly landscaped
civic square

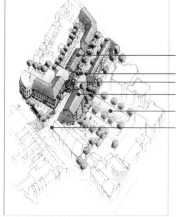

Attractive terraced housing
fronts traffic-calmed street
Library square
Mixed-use corner block
Flats over shop
Public car park in interior
Main civic square and bus stand

centre along one of Southampton's linear communities. This thinking can also be broadened out to address the surrounding catchment area in terms, for example, of improving the quality and attractiveness of walking and cycle routes, open spaces and play areas as well as reinforcing community identity through well designed signage and street furniture;

- making the most of the opportunities for new housing and other uses in the catchments around existing centres by developing at appropriate urban densities;
- grasping the opportunities to create new local centres in areas that are currently remote from local facilities and services. Figure 2.6 (4) shows how the nucleus of a local centre could be developed through the replanning of an adjacent but free-standing retail park to become an integrated piece of the urban fabric. Similar possibilities may arise though the planning of urban extensions that could be designed so as to serve both a new and an existing community (5).

Figure 2.6
Replanning a stand-alone retail park to create a new district centre

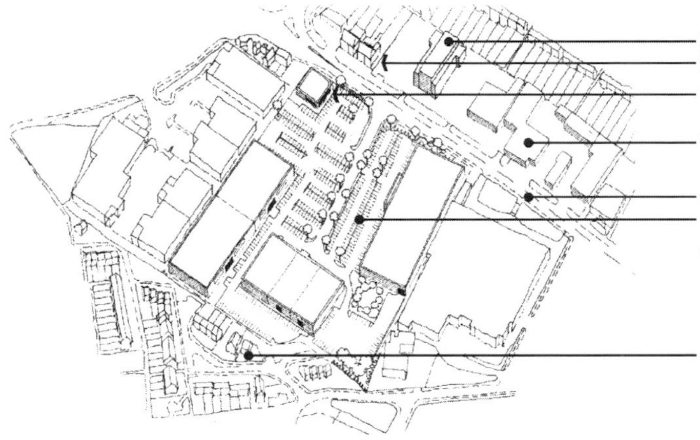

Disused cinema
Nucleus of local centre
Fast-food restaurant

Vacant industrial buildings

Main arterial
Retail warehouse estate and expansive parking lot

Disconnected housing estate

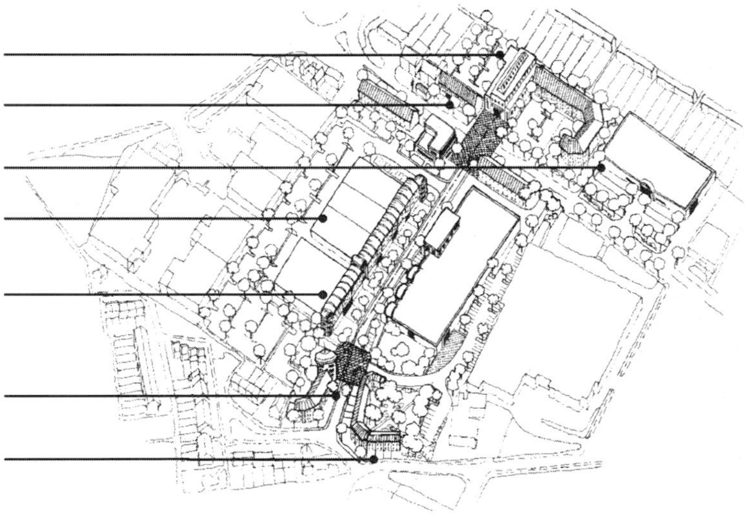

Conversion of derelict cinema to mixed-use block and square

Grain of existing frontage continued

Relocated retail warehouse

Restyled retail warehouses with offices above

South-facing flats over supermarket, meeting local shopping needs

New internal street connects residential neighbourhood and commercial centre

New housing knits together existing estate and local centre

2.5 Streets and street blocks

The development structure within the neighbourhood takes the form of street blocks defined by movement routes. These blocks are arranged in such a way as to enable direct pedestrian movement to and from important facilities and amenities including the centre and public-transport routes and stops. The pattern of routes is open-ended, providing a choice of routes to and from any given point. In general the street blocks become smaller closer to the centre to optimise pedestrian permeability.

Individual street blocks take the form of perimeter development with buildings facing outwards towards the edge of the block and the street. This provides the basis for active street frontage with windows and doors overlooking and opening out onto the street to provide good surveillance of the street and the activity within it. In residential areas rear gardens (whether private or shared) are enclosed by building frontage and walling.

There are no hard and fast rules to street and block dimensions. What is important is that both are designed in relation to their context and roles. This implies a variety of street widths from wide avenues to intimate mews courts. However, a number of important guidelines can be identified.

- Corners need careful design both to recognise their prominence at the junction of two routes and to ensure that they are turned with active frontage and continuity of passive surveillance of the public realm. Often this is more easily achieved with apartments, which can give greater scale and which do not require individual rear gardens which are difficult to accommodate within the corner.
- The orientation of blocks needs to be considered with solar potential in mind. A balance is needed here between seeking the optimum east–west alignment of streets to maximise solar potential and other design objectives, such as providing direct movement routes to local facilities and creating a robust perimeter block structure with continuity to the relationships between the fronts and backs of buildings. Variation in building heights and breaks in the building line can reduce shadowing and increase solar access within the block.
- Solar orientation will also influence thinking about the internal layout of the dwelling. For example, locating principal living rooms to maximise solar access could imply having kitchens and bathrooms to the front on one side of a street and to the rear on the other.
- Care needs to be taken over the design of the threshold between the street and the dwelling. Services, bins, cycles, space for home deliveries all need to be thought about here along with the need to balance surveillance of the street with privacy for the home.
- Streets should fulfil a variety of roles – a place to live, shop, park, drive, walk, cycle and catch the bus, and even, in quieter side streets, for children to play.

2.6   Optimising development density

Making efficient use of land and supporting local services by developing at appropriate urban and suburban densities is an important thrust of the new design approach. The term "appropriate" is deliberately used because development density will vary according to the location and accessibility of different sites, their context and setting, and the different types of housing being provided. Indeed, an important principle is that development density should be the outcome of a design approach that responds to these and other issues creatively rather than to a fixed design requirement.

Against this background a number of planning and design principles can be identified in seeking to create attractive and sustainable environments that make more efficient use of land and energy:

- Requirements for car parking can have a major impact on both the density of development and its quality. This is especially true of small sites of up to 1 ha where cars are parked on the surface. In some cases reducing the off-street requirement from two to one space per dwelling can increase site capacity by 50 per cent. Figure 2.7 (6) illustrates the differences in both townscape character and site capacity associated with different approaches to car-parking provision.

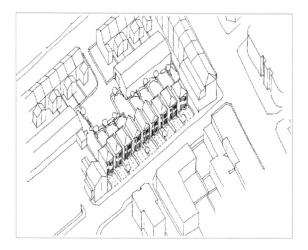
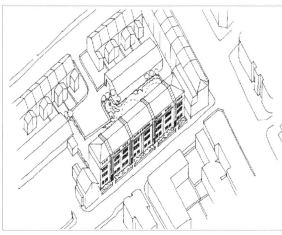

Figure 2.7
Removing the requirement for off-street
parking increases this site's capacity
from 10 houses to 32 apartments

- The efficiency of the street network and the arrangements for parking
  are critical factors on larger sites. For example, a traditional street
  grid with cars parked on-street can reduce the percentage site area
  required for roads and parking from 35/40 per cent to around 20
  per cent assuming a terraced housing form and around one parking
  space per dwelling.
- Thinking about requirements for new community facilities in relation
  to existing provision in the area is important. For example, instead of
  seeking further open space in an area that is already well served by
  open space it may be more beneficial to seek a financial
  contribution towards improving an existing space or play area or
  making better streets.

The logic of the walkable community again provides a useful starting point
in thinking about appropriate levels of density and car parking. The starting
point is to build up an understanding of how the site fits with its surrounding
context. For example, where are the local shops, schools and open spaces?
Where are the bus, tram and train stops and what destinations do they
serve? Where can you walk to in 5 minutes? What can be reached in 10
minutes?

Figure 2.8 (7) shows how this approach can be used to build up a picture
of the walkable neighbourhood around a 3 ha site in outer London. In this

Figure 2.8
Analysis of the facilities and services
within walking distances of the site
informs decisions on density car
parking and layout

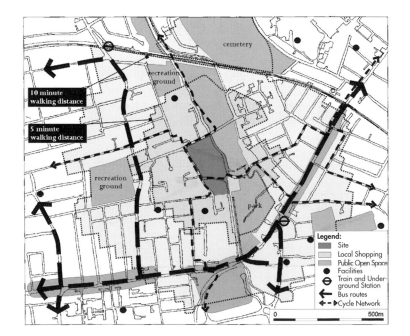

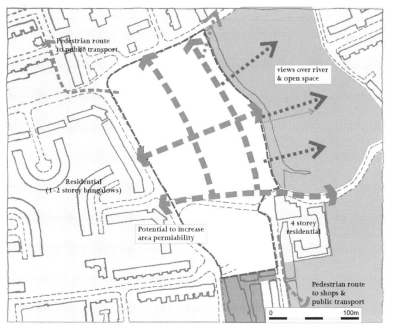

Figure 2.9
Redevelopment provides the opportunity to reconnect the surrounding community to local shops and open spaces

Labels on figure:
Pedestrian route to public transport
views over river & open space
Residential (1-2 storey bungalows)
Potential to increase area permiability
4 storey residential
Pedestrian route to shops & public transport
0   100m

case the analysis pointed to a level of accessibility to both facilities and public transport that was significantly better than that which would have been assumed from a cursory analysis.

This then provided the basis for proposing reduced levels of parking provision and no on-site provision of public open space. The analysis also helped in identifying pedestrian desire lines (see Figure 2.9) (8) from surrounding residential streets that had effectively been blocked by the site's former industrial use. These desire lines provided an important starting point in defining the structure for the site's redevelopment. The resulting development concept is shown in Figure 2.10 (9).

2.7   Some density rules of thumb

Development density will vary across an urban area reflecting different levels of proximity to the city, town and local centres, the public transport network and other facilities and amenities, such as open spaces and waterside areas. The matrix set out in Figure 2.11 (10) provides some guideline housing-density ranges derived from case-study design analysis of densities achieved on seventy sites across London that were designed with different mixes of houses and apartments and different levels of car parking. It indicates a very wide range of densities from 30 dwellings per hectare (dph) in areas away from facilities and public transport right through to over 400 dph in the very accessible city centre.

The foregoing discussion has been concerned with what planners term "net housing density", that is, the ratio of dwellings (or habitable rooms or floorspace) to an area of land developed primarily for housing. But, clearly, towns and cities need to provide a range of other uses, such as shopping, employment and leisure, which need to be either within walking distance of home or accessible by public transport. A "town" or "district" density includes allowance for these non-housing uses and should be much lower than the net housing density. The important principle is that moderate to high net housing densities are supported by good-quality local facilities, including parks and open spaces. This produces lower "town or district" densities but maximises the proportion of the population who can access local facilities and amenities on foot (11).

Figure 2.10
A mix of town houses and apartments achieves a density of 110 dwellings per hectare with 0.6 parking spaces per dwelling

Figure 2.11
A matrix giving guideline density ranges for sites with different levels of accessibility to local facilities and public transport

Key

Abbreviations

hr	= habitable rooms
u	= unit
ha	= hectare
U/ha	= dwelling units per hectare
Hr/ha	= habitable rooms per hectare
Ave	= average number of habitable rooms per dwelling

Definition of site setting

Central	= very dense development, large building footprints and buildings of 4–6 stories and above, e.g. larger town centres and much of Central London
Urban	= dense development with a mix of different uses and buildings of 3–4 stories, e.g. town centres, along main arterial routes, and substantial parts of Inner London
Suburban	= lower-density development, predominantly residential buildings of 2–3 stories, e.g. some parts of Inner London, much of Outer London

			Option 1	Option 2	Option 3
Car Parking Provision			High 2–1.5 spaces per unit	Moderate 1.5–1 space per unit	Low less than 1 space per unit
Predominant housing type			**Detached & linked houses**	**Terraced houses & flats**	**Mostly flats**
Location		**Setting**			
Site within town centre 'ped–shed'	6 ↑ Accessibility index ↓ 4	Central			240–1100 hr / ha 240–435 u / ha Ave. 2.7 hr / u
		Urban		200–450 hr / ha 55–175 u / ha Ave. 3.1 hr / u	450–700 hr / ha 165–275 u / ha Ave. 2.7 hr / u
		Suburban		240–250 hr / ha 35–60 u / ha Ave. 4.2 hr / u	250–350 hr / ha 80–120 u / ha Ave. 3.0 hr / u
Sites along transport corridors & sites close to a town centre 'ped–shed'	3 ↑ ↓ 2	Urban		200–300 hr / ha 50–110 u / ha Ave. 3.7 hr / u	300–450 hr / ha 100–150 u / ha Ave. 3.0 hr / u
		Suburban	150–200 hr / ha 30–50 u / ha Ave. 4.6 hr / u	200–250 hr / ha 50–80 u / ha Ave. 3.8 hr / u	
Currently remote sites	2 ↑ ↓ 1	Suburban	150–200 hr / ha 30–65 u / ha Ave. 4.4 hr / u		

hr = habitable rooms
u = unit
ha = hectare

2.8   Some broader issues and key points

The discussion in this chapter has been largely concerned with the structure of the urban environment. This is a fundamentally important component in planning for a more sustainable future, but in itself will not deliver sustainability. That depends also on more detailed design considerations, explored in subsequent chapters of this book, but it also turns on a number of broader issues. These include the following:

- The creation of socially mixed and inclusive communities. One of the lessons of much post-war housing is that the split between market and social provision has helped to polarise society into ghettos of rich and poor, imposing incalculable economic and social costs on future generations. The lesson must be about creating a wider mix of housing opportunity and choice and avoiding concentrations of particular housing types and tenures.
- The provision of services and facilities that meet a range of needs. Pristine out-of-town shopping centres may be efficient places for those with access to a car to shop, but they lack the character, serendipity and community focus of a town centre or high street with its mix of high-fashion, second-hand, ethnic and convenience stores, pubs and access for all.
- The engaging of local communities in discussion about how they see their neighbourhood and their priorities and aspirations for the future. The dialogue should be honest, open, ongoing and with a real commitment to changing plans and designs to reflect people's views.
- The provision of quality public transport services. This is a fundamental prerequisite in reducing reliance on the car. The next chapter sets out the key challenges and potential in seeking to turn around decades of under investment in public transport.

- The delivery of excellent local facilities and services. If people are to walk to local amenities then these must meet their needs, as well as any to which people can drive. This applies to schools, open spaces and play areas, local shopping facilities, as well as the walking routes to and between these facilities.
- The recognition that long-term management and maintenance are as important as the initial design. New development must be designed with management and maintenance in mind, not just in terms of the choice of materials and landscape but also with a clear definition of who will be responsible for what and a commitment to pay for maintenance over the long term.
- The vision of new development as a catalyst for the improvement of existing areas. This demands excellent design, but it could also include a local "community chest" to pool contributions from a range of development projects to be spent on local community projects. This could help encourage a more positive public attitude to development and change, which becomes increasingly important the more development moves from a few large sites to a multitude of improvements to the physical environment and community facilities. Such approaches also increase in importance as more emphasis is placed on the reuse of very small sites.

It must be clear from this wide range of issues that sustainable planning and design can belong to no one discipline. Success requires a holistic and integrated design approach that draws on skills in planning, urban design, architecture, landscape design, building and services engineering, community consultation and development, and much more besides.

REFERENCES

1. The Urban Task Force (1999), Towards an Urban Renaissance: final report of the Urban Task Force. E&FN Spon, London.

2. See also Reference 1, p. 66.

3. Government Office for the South East (1998), Sustainable Residential Quality in the South East. Government Office for the South East, Guildford.

4. Ibid.

5. The Prince's Foundation et al. (2000), Sustainable Urban Extensions: planned through design. The Prince's Foundation, London.

6. London Planning Advisory Committee et al. (1998), Sustainable Residential Quality: new approaches to urban living. Greater London Authority, London.

7. London Planning Advisory Committee (2000), Sustainable Residential Quality: exploring the housing potential of large sites. Greater London Authority, London.

8. Ibid.

9. Ibid.

10. Ibid.

11. Department of Environment, Transport and the Regions (1998), The Use of Density in Urban Planning. DETR, London.

FURTHER READING

Department of Environment, Transport and the Regions (1998), Places, Streets and Movement: a companion guide to Design Bulletin 32. DETR, London.

Department of Transport, Local Government and the Regions/Commission for Architecture and the Built Environment (2000), By Design: urban design in the planning system – towards better practice. Thomas Telford Publishing, London.

Department of Transport, Local Government and the Regions/Commission for Architecture and the Built Environment (2001), Better Places to Live: by design. Thomas Telford Publishing, London.

English Partnerships and Housing Corporation (2000), The Urban Design Compendium. English Partnerships, London.

Rudlin D. and Falk N. (1999), Building the 21st Century Home: the Sustainable Urban Neighbourhood. Butterworth-Heinemann, Oxford.

Urban Villages Group (1992), Urban Villages. Urban Villages Group, London.

3

Transportation

Robert Thorne and William Filmer-Sankey

3.1 Energy use and transport

Travel is a natural human urge; it keeps us fit, broadens our experience and brings new contacts. It is also essential for moving food and supplies to maintain the fabric of urban life. Travel, furthermore, is not just a means, but an end in its own right. The act of travelling is in many cases as important as reaching the destination.

Travel is, in short, both a sustainable and a sustaining activity; cities rely on it. The problem of recent years is that the means of travel – transport – has come to dominate our lives at the expense of the journey. Transport, of people and of goods, now accounts for a significant proportion of the UK's energy use. The most popular form of transport – the motor car – is the least energy efficient (1). The distance we travel for essential journeys (to work, school and shops) has increased by up to 40 per cent in the past twenty years alone. And the journey takes longer: the average speed of traffic in Central London (10 mph/6.25 km/h) is now as slow as it was in the nineteenth century.

As journeys become longer, slower and more wasteful of energy, so too have they ceased to fulfil their enriching and broadening role. Travel by car (which accounts for 60 per cent of journeys) gives virtually no physical exercise when compared with walking or cycling (which also extend life expectancy) (see Table 3.1). It increases levels of mental stress. Opportunities for conversation are limited. Current transport methods thus positively discourage human interaction, which is itself the bedrock of urban living. Such transportation restricts interaction with the built environment for those in cars or adjacent to them; it discriminates against the one in three people who do not own a car, and it kills an unacceptably high number of people, whether by knocking them down or by producing health-damaging

Table 3.1
Nature and scale of some impacts of transport on health in London (2)

Impact	Scale
Road accidents, fatalities (2000)	286
Road accidents, total casualties (2000)	46,003
Percentage considering noise from road traffic a nuisance (GB figure, 1991)	63
Calories consumed for 70 kg person (kcal/h), driving a car	80
Calories consumed for 70 kg person (kcal/h), walking at 5 km/h	260
Calories consumed for 70 kg person (kcal/h), walking at 7 km/h (brisk walk)	420
Estimated net life extension, compared to whole population, of those who walk or cycle to work	2 years

Source: Informing transport health impact assessment in London, AEA Technology/NHS Executive London, 2000, and TfL, 2001.

Figure 3.1
Traffic jams in London: the sign of a
sick city

pollutants (3) (see also Appendix C). If transport is unsustainable, it makes the city sick too (see Figure 3.1).

3.2 Defining sustainable transport

Sustainable transport can be defined in a number of ways. In the first place it should "meet the needs of the present without compromising the ability of future generations to meet their own needs" (4). More immediately, it should promote human health, both mental and physical, and provide the opportunity for social interaction and enriching urban experiences.

Three changes are required to achieve these goals. The first is a reduction in the need to travel, and the distance that people have to travel, particularly for those essential journeys to work, school and shops, which account for 46 per cent of journeys of more than 1 mile (1.6 km) (5). Since 70 per cent of these "essential journeys" are currently made by car, the potential for reduction is clearly considerable. Goods as well as people need to travel less, particularly food. The current fashion for organic food has deep contradictions: a recent study calculated that a shopping basket of twenty-six organic products could have travelled 241,000 km and released as much CO_2 into the atmosphere as does the cooking of meals in an average four-bedroom house for eight months (6).

Second, we need to change our mode of travel. For people, this means changing from cars to foot or cycle for short journeys and to public transport for longer journeys. Even in existing circumstances, such a change, particularly to cycle, can already result in savings on journey times. Research in London shows that cycling is now quicker than car for a variety of journeys: for example a 1.7 mile (2.7 km) journey in the centre takes under 20 minutes by cycle but 30 minutes by car or public transport (7). Third, and finally, we need to make cars more energy efficient and less polluting.

In making these changes, we must not see transportation in isolation. As this book demonstrates, many of the issues affecting it have a wider relevance; solutions that work in one area may be counterproductive in others. A sustainable transport system must both feed into and be informed by the wider needs of the city.

The debate on sustainable transport is often dominated by the application of new technology to motor cars and there is no doubt that this has an important role to play (for example through the increased use of electricity or hydrogen fuel cells) in reducing the environmental impact of motorised transport. But it will not have any impact on the other, equally serious, disadvantages of existing travel patterns: people will still have to travel ever-longer distances at ever-slower speeds and be denied the opportunities for social interaction. Indeed, it can be argued that the best hope for achieving sustainable transport lies not in technological tinkering with the motor car but with a radical shift in land-use planning, reflecting in turn changing social patterns (such as home working). The wider implications of this were discussed in the previous chapter, but it is worth considering its specific potential to encourage more sustainable urban transport.

3.3 Urban design priorities

It is land-use policies, reflecting perceptions of the past 100 years, that have shaped today's urban development and, in particular, patterns of transport. When land was available, and cities seen as overcrowded and unhealthy, planning (from the Garden City movement onwards) encouraged dispersal: people should move away from the old centres to new, low-density settlements on the edges.

The growing popularity of the motor car in the second half of the twentieth century reinforced the dispersal of human activity from the centre to the suburb. Its engineering demands came increasingly to dominate urban (and suburban) planning and the layout of new settlements. It also introduced a further separation – between dwellings and shops – as people found it easier to buy all their goods in one place and load them into their car, and as shops became superstores, confined to town edges by the need to provide massive car-parks. We can now see clearly that the result of these movements was not the hoped-for liberation but rather the increased pollution, congestion, time spent travelling and the decay of city centres. But in the UK we are still saddled with the results of planning policies drawn up around the car. This is true even in those cities that were laid out around a public-transport grid of metros and tramlines. The grid often survives (and can be revitalised) but it is swamped by the twentieth-century obsession with the narrow requirements of the motor vehicle.

It is these planning policies that need, quite literally, to be reversed. For transport, as recent UK Government guidance has emphasised, the major issues are maximising development density in locations well placed for public transport, and shortening journey lengths by mixing home, work, school, shopping and leisure facilities in the same area (8).

The relationship between density and transport is key. On the one hand is parking provision. The average parking space takes up $11.52\,m^2$. The higher the level of car ownership, the greater the need for parking, the greater the environmental footprint of the new development and the lower its density. Car Clubs, now working successfully in several UK cities (including Bristol and Oxford), can make a significant difference by providing a pool of cars that members can use when they need to travel by car (9). A single Car Club vehicle can satisfy the needs of up to twelve individuals who would otherwise need one of their own, thus freeing valuable space for other uses.

On the other hand is the provision of viable public transport. Density is a tool to ensure that viability: the higher the density, the better the level of service that can be provided (see Figure 3.3 and Table 3.2). Public-transport services that provide a genuine alternative to the private car will only work where there are sufficient people: hence the need for clusters of higher density within walking distance of public-transport stops and interchanges. The precise density and the catchment area will vary with the type of public transport, but at least it should be designed to justify the provision of a regular bus service.

The segregation of living, working and shopping is another trend that needs to be reversed. The mixing of uses (residential, work, retail) in the same area is one of the most obvious ways of reducing the current average distances travelled. In centres, where densities are high, the mix of use can

Figure 3.2
A classic cul-de-sac: low-density twentieth-century suburban sprawl

Figure 3.3
Freiburg-im-Breisgau, Germany

Table 3.2
Catchment population for public transport (10)

	Minibus	Bus	Guided bus	Light rail	Rail
Stop interval	200 m	200 m	300 m	600 m	1,000 m+
Corridor width/ area served	800 m	800 m	800 m	1,000 m	2,000 m+
Catchment per stop	320– 640	480– 1,760	1,680– 3,120	4,800– 9,000	24,000+

Figure 3.4
Mixed-use development in Fulham,
London

Figure 3.5
The bicycle-hire store at Freiburg-im-
Breisgau, Germany

be achieved vertically by having, for example, shops at ground-floor level, offices above and flats at the top (see Figure 3.4). Such a mix ensures that the area always has life and movement.

Further out, horizontal mixed-use development provides zones of different activity: the parade of shops serving a surrounding block of purely residential streets is the classic Victorian and Edwardian pre-car suburban pattern. Mixing uses in these ways ensures that a range of services is within a reasonable distance (see Table 3.3), thus encouraging cycling or walking and giving new opportunities for social contact and interaction. Locating major centres of working and shopping near to public transport interchanges and connecting them into the local cycle or footpath network will help to persuade people to leave their cars at home (11).

The final essential is, self-evidently, a public-transport system that is integrated, smart, safe and efficient. At Freiburg railway station, for example, it is possible to hire a bicycle to use during your visit to the city (see Figure 3.5).

In short, if you want to maximise sustainable transport, new developments should be based on propinquity, not on dispersal. They will have a very different layout to conventional, car-based designs. They will put people first by creating "walkable neighbourhoods", in which it should not be necessary to resort to your car for the normal daily needs of life and which can be laid out principally for pedestrians and cyclists. For instance, the new community at Osbaldwick, York, is planned to fit within a 5-minute walking circle (see Figure 3.6).

Table 3.3
Creating a walkable neighbourhood: all "local hubs" should be within easy walking and cycling distance (12)

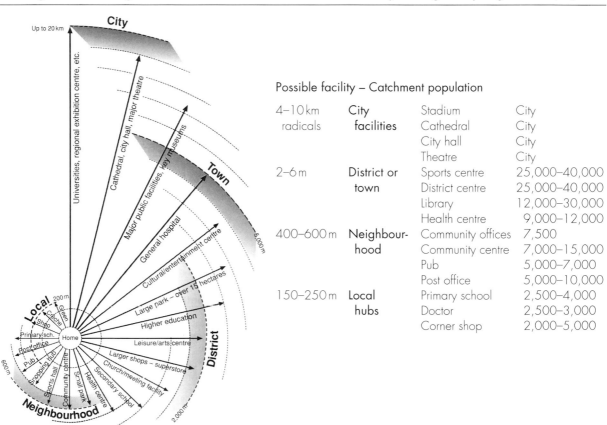

Possible facility – Catchment population

4–10 km radicals	City facilities	Stadium	City
		Cathedral	City
		City hall	City
		Theatre	City
2–6 m	District or town	Sports centre	25,000–40,000
		District centre	25,000–40,000
		Library	12,000–30,000
		Health centre	9,000–12,000
400–600 m	Neighbour-hood	Community offices	7,500
		Community centre	7,000–15,000
		Pub	5,000–7,000
		Post office	5,000–10,000
150–250 m	Local hubs	Primary school	2,500–4,000
		Doctor	2,500–3,000
		Corner shop	2,000–5,000

This chart is indicative and is based upon city-scale urban areas. Catchments will vary in specific areas.

3.4 Detailed design

Built-in flexibility is essential. A problem of the previous trends in land-use planning is that the underlying ideas proved temporary, but the urban designs they produced live on. An important element of designing for sustainability is that it is not bound by dogma, but can be adapted to meet future demands. This is particularly true of roads and services: houses scattered around a cul-de-sac can easily be demolished, but the infrastructure of low-density building, block size or street pattern is far harder to replace.

This challenge is reflected in the crucial issue of the future of motor vehicles in sustainable cities. The continuing role of the car cannot be denied, particularly in long-distance and leisure travel and in linking the countryside to the city. But should it be allowed into and through the city? The building of a bypass, by taking traffic away from a town centre, may erode its life and activity. "Walkable neighbourhoods" are fine but we need to beware of creating ones that turn their backs on their surroundings. We rightly frown on cul-de-sacs as being dead-ends; a dead-end settlement is just as dangerous. The handling of routes through places is a design challenge that must be met head-on, and not diverted around the edge.

Design has an important role to play in making places feel good for pedestrians and cyclists. Block sizes can be designed to give a balance between ease of access, mix of uses and privacy, while the careful laying out of grids and taking care over connections between neighbourhoods ensures that routes for pedestrians and cyclists are as direct as possible. Street widths, the treatment of different types of street, and frontages contribute to the variety of experience that those on foot or cycle can notice and appreciate, but those in cars cannot. The current trend of nomenclature and definitions of roads reflects their subservience to the needs of motor cars. Changing those descriptions is more than a symbolic gesture: it reflects a changed perception of what – or rather who – streets are for (see Table 3.4).

Cyclists and pedestrians have very practical needs, which must be addressed by good design. An obvious example is the need for conveniently placed, secure cycle storage in the home: cycles that have to be retrieved from a cupboard and carried down several flights of stairs will not get used. The design of spaces for the young and the visually and mobility impaired is equally important to ensure that access is genuinely inclusive.

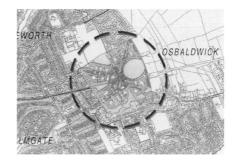

Figure 3.6
Osbaldwick, York

Table 3.4
Redefined street types (13)

Conventional capacity-based terminology	Streets that combine capacity and character
Primary distributor	Main road Routes providing connections across the city
District distributor	Avenue or Boulevard Formal, generous landscaping
Local distributor	High Street Mixed uses, active frontages
Access road	Street or Square Mainly residential, building lines encouraging traffic calming
Cul-de-sac	Mews/Courtyard Shared space for parking and other uses

Figure 3.7
Careful design of public-transport
facilities, Walsall

Far greater priority needs to be given to the design of public transport. In practical terms, this means ensuring that it functions well and is accessible to all (including the mobility or visually impaired). But there is the challenge of establishing status and image: the design of stops, stations and interchanges must counteract the prevailing impression that public transport is only for those who cannot afford a car by being made to look smart and safe. Their location, their relationship to the urban grid and their architecture must all be used to emphasise the status of public transport, its central role at the heart of the place and its ownership by the community (see Figure 3.7).

3.5   Future technologies

Technology is usually thought of in terms of improvements to fuel efficiency and cleanliness, but it is also vital to achieving the new urban-design priorities. For public transport, there are new Light Rail Transport (LRT) and guided bus systems, which can often make use of the existing nineteenth-century urban grid, originally laid out with trams in mind. Of equal importance is the use of satellite-tracking and second-generation mobile phones to provide real-time information about the arrival times of buses, for example. Nothing contributes more to the sense of second-class citizenship of public-transport users than the long wait at a cold bus stop. It also gives a sense of powerlessness, in contrast to car users, who can move as and when they wish. The technology to provide real-time information, not just at stops, but to homes, shops and mobile phones, already exists.

The impact of the internet on shopping patterns is potentially of huge importance to transport patterns, given that 19 per cent of journeys over 1 mile/1.6 km are currently made to shops and given the efforts being made by the major supermarkets to offer internet-ordering and home-delivery services. There is the potential for a significant reduction in the number of car journeys if this service catches on for deliveries of basic groceries. However, it runs the risk of reinforcing social divisions (as car ownership currently does) if the internet is not made as widely accessible as possible. Furthermore, it depends on new developments having built-in facilities to receive orders at all times. This can take the form of special storage in each dwelling or of a "corner shop" to which deliveries are made from a central place and from which customers can collect them. Such a corner shop could become a focus of movement and activity, attracting other services to it.

3.6   What can be done now and in the future?

It is a curious fact that damaging changes in land-use planning, such as the explosion of out-of-town shopping centres, can happen with frightening speed, but that measures to reverse the results can only be implemented over a long period. Testing new technologies, too, is something that cannot happen overnight. New transport infrastructure needs to go through a long planning process and inevitably takes time to commission and build. Making major changes to the inherited, car-dominated layouts of so many of our cities is even harder. It all requires a long-term commitment from Government, Local Planning Authorities and, of course, voters. In order to maintain the current spirit of hope and optimism, it is essential that, in parallel with punitive measures to reduce car use, progress is seen to be made in promoting sustainable alternatives. Effective "early wins" can be achieved in a number of areas.

Figure 3.8
Human interaction at the school gates,
Ealing, London

- Improvements to the street environment, for instance:
  - reducing traffic speeds and giving priority to pedestrians and cyclists, in centres and in Home Zones, 20 mph/30 kmh

Zones and Safe Routes to Schools (see Figure 3.8) (14). It is very important that our children are encouraged to get used to cycling and walking, as leisure activities as well as means of transport; it is their future;

- creating and maintaining an attractive public realm, which encourages people to enjoy their journey, be aware of their surroundings, and to stop and talk. Improved street maintenance, and a more co-ordinated approach to street design, requires no new legislation and can have an immediate impact on how people chose to travel.

- Creating better pedestrian and cycle links, which are not simply a cobbling together of existing fragments of roads but form co-ordinated, distinctive and (above all) useful networks. Cyclists and pedestrians must not be made to feel inferior to car users.
- Improving the image of public transport through investment in well positioned and designed stops, accessible stations and real-time information.
- Ensuring that new developments:
  - are located close to public transport routes and interchanges, and have the right density to maximise the use of public transport and minimise the need for car ownership;
  - contain a good mix of uses, designed to be easily accessible on foot and by cycle;
  - are designed with good cycle storage, easily accessible internet facilities and a place to receive and store deliveries;
  - are flexible enough to adapt to changing needs.

Such improvements will be small, and will not immediately show up in the statistics. But they are important as they will show the positive human benefits, the "poetry, optimism and delight", of sustainable transport. For it is through this route that the individuals who make up the statistics will themselves realise its many benefits.

3.7   Guidelines

1. Travel is not a means but an end in its own right: it should be both sustaining and sustainable.
2. Sustainable transport means three things:
   - reducing people's need to travel, both in the number and length of journeys;
   - changing the normal mode of transport, from motorcars to foot and cycle;
   - making motor vehicles more energy efficient and less polluting.
3. The key to sustainable transport lies not so much in new technology as in changed urban design priorities such as:
   - higher-density development, located around public-transport stops and interchanges;
   - mixing of uses (houses, shops, workplaces, schools and public facilities) within the same area;
   - better public transport.
4. Good design has a vital role to play in reversing not just the reality but also the perception that public transport is only for the second-class citizen.
5. Sustainable transport cannot be achieved instantly. But "early wins", such as street improvements and better public transport, can show its potential to improve everyone's quality of life.

REFERENCES

1. Bannister, D. (2000), The tip of the iceberg: leisure and air travel. Built Environment, volume 26, number 3, p. 229, table 3 – energy in kWh per passenger-km: car, 0.58; rail, 0.44; bus, 0.26.

2. Greater London Authority (2001), The Mayor's Transport Strategy, fig. 2.11.

3. Royal Commission on Environmental Pollution (1997), Twentieth Report: transport and the environment – developments since 1994, pp. 19–20.

4. World Commission on Environment and Development (1997), quoted by Department of the Environment (1997), Planning Policy Guidance Note 1 (Revised): general policy and principles, p. 3, para. 4.

5. Reference 3, table 4.2.

6. Sustain/Elm Farm Research Centre (December 2001), Eating Oil: food supply in a changing climate.

7. Reference 2, p. 52, fig. 2.21.

8. Department of the Environment, Transport and the Regions (March 2001), Planning Policy Guidance Note 13: transport. See also, Department of the Environment, Transport and the Regions (October 1998), Planning for Sustainable Development: towards better practice, especially chapter 2 (Realising the potential of existing urban areas) and chapter 6 (Incorporating other sustainability issues: parking).

9. Community Car Share Network (summer 2001), The Road Ahead: community car share network newsletter, issue 4.

10. English Partnerships and The Housing Corporation (2000), The Urban Design Compendium, table 4.1.

11. As reference 8.

12. Towards an Urban Renaissance: final report of the Urban Task Force, fig. 1.3.

13. Reference 10, table 4.2.

14. For Home Zones, see Biddulph, M. (2001), Home Zones: a planning and design handbook. The Policy Press. For 20 mph zones, see Slower Speeds Initiative (2001), Killing Speed: a good practice guide to speed management. Also Department of the Transport Traffic Advisory Leaflet 9/99 (1999), 20 mph Speed Limits and Zones. For Safe Routes to Schools, see Department of the Environment, Transport and the Regions (June 1999), School Travel Strategies and Plans: a best practice guide for local authorities.

FURTHER READING

Barton, H., Davis, G. and Guise, R. (1995), Sustainable Settlements: a guide for planners, designers and developers.

DTLR and Cabe (September 2001), Better Places to Live: by design.

DETR (September 1998), Places, Streets and Movement: a companion guide to Design Bulletin 32, residential roads and footpaths.

Landscape and nature in the city

Christina von Borcke

4.1 Introduction

Landscape design and the presence of nature are critical to the quality of our urban environment. Their role is not only to make places look greener, but also fundamentally to influence the form of development.

Landscape plays a profound role in the development process. It is not an add-on, but rather forms the basis for creating places. Landscape is not only trees, shrubs and lawn, added for their aesthetic value. Rather, landscape combines landform, ecosystems and open-space networks that form the natural environment and sustain planting. There is a tendency to think of the city and nature as diametrically opposed (see Figure 4.1). In this context, development is seen as an imposition on natural surroundings. Development, though, is built on land and landform that is unique to its location, with its own natural landscape and its intrinsic sense of place.

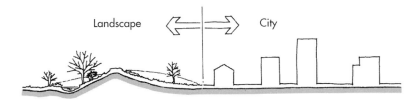

Figure 4.1
The misconception of landscape and city as diametrically opposed uses that cannot be brought together

Recognising this simple fact brings a completely different approach to the roles of landscape in the city, in new developments and in the design process. It suggests that landscape is integral to city living and needs to be considered as a central part of any development activity. It further suggests that the city is added to the natural setting and responds to it rather than the other way around (see Figure 4.2). Landscape is a fundamental element of the design process, and may even be the starting point of design. Shouldn't the city grow from its setting rather than be imposed on it? Should the land not then inform a unique and site-specific response to development rather than be subjected to the application of rigid national standards?

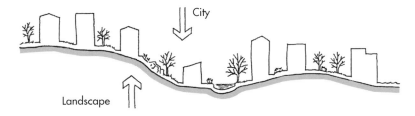

Figure 4.2
Landscape as grounding and context from which the city has developed

The benefits of this approach are at the core of the sustainability agenda and especially sustainable urban design. The approach builds on a balanced coexistence of city and nature, which has benefits for both. Nature not only holds aesthetic value in the city context but also improves the local microclimate, relieves environmental pressures on the city region and provides mental relief and contrast for urban dwellers. Nature and landscape in the city are hence important for improving the quality of life in urban areas and making those areas more sustainable in every sense of the word – ecologically, socially and economically:

ecologically – affecting microclimate, creating wildlife habitats;
socially – making places more likeable, hence heightening the sense of ownership, counteracting urban stress, improving quality of life;
economically – retaining property values because of a perceived better quality of life.

All these contribute to a more durable development with a longer life span.

Spaces between buildings form the basis of our landscape – these are areas where the ground is exposed and where plants may grow and water may penetrate the ground. The following discussion will highlight how landscape can contribute to the creation of more sustainable urban areas in strategic terms as well as in detail.

4.2 The wide-ranging benefits of landscape in the city

When talking about landscape in the city, one should not forget the primary role of urban areas – they provide the main locations for human habitation and interaction. They are places where people live, work and conduct their daily lives. The sustainability debate recognises that and pushes the point even further by suggesting that compact cities can lead to a more sustainable society. Living at higher densities will limit travelling distances and hence use less energy and fewer resources for transport. The reality, though, is that many people choose to live outside of cities even when otherwise depending on them for work and facilities. Research has shown that there are two main reasons for this trend: the lure of living in a physically attractive environment, and the search for a different type of community and lifestyle (1). These are environments that offer more space in the home as well as in the public domain, such as parks, fields and open spaces. Providing spacious modern dwellings, together with more public space in cities in the form of urban parks of different character and sizes that provide pleasant green views and contrast, can help to make cities more desirable places to live for a wider range of people. It can change the perception of the city and improve the quality of life that cities offer.

The role and benefits of landscape in the city are multifold. They affect:

- the *microclimate* of the city, making it a more pleasant place to live;
- *people's perception and quality of life* in the city, creating more green spaces of varied nature with more trees and greener views;
- the *diversity of species* that can live in the city, creating a real and diverse ecosystem with a range of habitats for wildlife.

Microclimate

It has long been understood that vegetation of any form influences the microclimate around it.

Vegetation is known to:

- draw CO_2 from the air through the process of photosynthesis and produce oxygen in turn (for example, over 99 years, five trees take up as much CO_2 as is produced by driving one family car for 1 year) (2);
- bind airborne particles, therefore reducing dust (foliage acts as an impingement filter, trapping airborne particles until they are washed away by rain) (3);

- bind numerous urban pollutants;
- absorb noise;
- raise local humidity by absorbing rainwater and/or delaying it entering the drainage system;
- make the ambient temperature more temperate (lowering it in summer, raising it in winter);
- act as a windbreaker, reducing wind speeds by up to 50 per cent and hence reducing the associated wind-chill factor – this in turn can lower the heating requirements for buildings (4).

Elaborating on all these processes and their benefits is not possible in the space available, but it is important to understand that vegetation positively influences climatic conditions (Appendix F provides more detail). It should therefore be better integrated into the design and layout of new urban and suburban areas.

People's perception and quality of life

The effect of vegetation on our environment in turn has a positive effect on human health through cleaner air, fewer airborne particles and a better microclimate. But there is a further dimension. Landscape in the city, such as parks, gardens, planting and especially trees, can make urban areas more attractive to live in and can raise the quality of urban life. Looking at the factors that have drawn people out of urban areas to live in the suburbs can give us clues as to what is missing in the city – namely greener environments that contrast with the hard lines and rigidity of buildings.

The influence of landscape goes further than improving the aesthetic value and ecology of cities. The existence of landscape in the city can influence the human psyche and well-being. Clinical trials have shown that hospital patients looking out over trees have a faster recovery rate, lower blood pressure and need less medication than patients who look out over paved areas (5). If something this simple can help the sick recuperate, lessons can be learned to improve our daily environment.

The presence of trees and mature landscape can to some degree also influence human behaviour. There is anecdotal evidence that people living in tree-lined streets are less prone to show violent behaviour, be depressed or abuse drugs (6). Trees along streets can influence driving behaviour. People drive more calmly and are more alert. This is not to suggest that our physical environment can solve all our social problems, but it can help stimulate more positive responses. Even if further research is needed to substantiate these claims, everybody agrees that avenue planting visually improves the quality of the street by deflecting attention from cars and should therefore be taken more seriously in the design process (see Figure 4.3).

But why does the existence of planting or the view over landscaped areas affect our well-being? Could it simply be colour? According to colour therapy, green is the neutral colour in the middle of the visible spectrum (between red, orange and blue, violet). The lens of our eyes focuses green light exactly onto the retina. Minimal muscular adjustment is needed (7). Green is therefore the easiest colour for the eye to see and functions like a "tonic" when exhausted. It also soothes and comforts the mind when tired and weary. Craftsmen in ancient Greece used this knowledge by using a green byrol (transparent material) to rest their eyes, while today the majority of sunglasses have a green tint to do the same thing. It is not without reason that the colour green represents peace and balance – an attribute currently not linked to cities.

Incorporating such landscape as parks, trees and general planting into the city can therefore help to counteract visually the stress created and

Figure 4.3
Avenue planting creates a more pleasant environment on the street and deflects attention from parked cars

experienced in cities as a result of traffic, density and noise. These landscapes do not have to be large parks but can also be integrated as small greens and pocket parks within the urban fabric. Green vistas can calm our senses and reduce tension. The urban environment becomes more pleasant and some of the reasons why people move out of cities are addressed.

Diversity of species

Plants are a source of food and shelter for insects, birds and small mammals. These in turn are the food sources for other animals. Plants need soil, water and light to grow and sustain this simplified food chain. But not any planting will do. A diversity of plant species is needed to attract different insects and birds. Native species are generally more attractive to local wildlife than exotic species as their fruits and flowers are more palatable to wildlife (8). This does not exclude the use of exotic species per se. Often they are very useful for their striking shape and/or ability to grow in stressed and polluted environments. But judgement needs to be made on the main aim when using planting.

Vegetation is the key to creating and sustaining biodiversity. But it is not one plant or tree species by itself that can attract wildlife. Rather it is the combination and, especially, diversity of species that forms the attraction. The combination of plants used is dependent on the type of soil, the availability of water and the existing climate. All these together form the local habitat that is attractive to different birds, insects and small animals. Creating a large number of different but interlinked habitats (such as forests, open tree planting, hedgerows, meadows, grassland, ponds, riverbanks and streams) gives the best chance of attracting wildlife and improving the biodiversity of the city.

Not all plants, though, will grow in all situations or locations. Some are better for dry soils or wet locations, sunny or shady, exposed or protected. Care needs to be taken when selecting plants for specific extreme situations. In general, one can assume that local species are better adapted to local climatic conditions and therefore need less care and water. The table in Appendix F lists a large variety of plants with their suitability for different purposes or situations. Pollution-resistant plants are well suited as street trees, while plants attracting insects and birds are best suited for open spaces and gardens.

4.3   From landscape strategy to detailed implementation

In the context of making places more sustainable, one needs to address the issues at different scales. Some fundamental principles need to be put in place at a strategic level so that detailed solutions can be effective. Detailed solutions strengthen the effects of the strategic principles. This general statement is also true for making landscapes more sustainable and, through its role, lead to more sustainable urban design. The slogan "think globally – act locally" encapsulates the relationship between strategy and detail.

Relating strategy and detail to sustainable landscape design is fairly straightforward. There needs to be enough space for nature and landscape elements to grow in the first place before we can select the most appropriate and beneficial plants for the desired effect. To increase the biodiversity in the city, open-space networks are advantageous in that they can form wildlife corridors that help a variety of animals to pass through or move deep within our urban areas (see Figure 4.4) (9). With these corridors in place the potential increase in biodiversity is far higher than with isolated

Figure 4.4
Open-space networks as wildlife corridors in urban areas

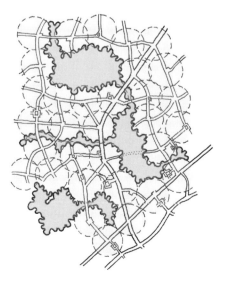

pockets of open space. The type of soil, access to water and selection of plants further influence the biodiversity of these places in detail. On the city scale, these corridors and networks are made up of parks, back gardens, allotments, trees and hedges but more importantly of leftover land, riverbanks, lakes, railway cuts, green roofs, climbers and buffer planting.

A landscape strategy outlining landscape corridors and important linkages on a city scale can safeguard land from future development and hence retain its ecological function as well as providing improved access to open spaces. The city of Hamburg has embedded such a landscape strategy in its planning system. Landscape corridors stretch from all directions deep into the city centre (see Figure 4.5) (10).

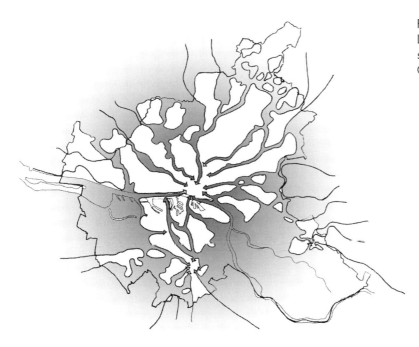

Figure 4.5
Landschaftsachsenmodell/Landscape strategy for the city of Hamburg, Germany

The principle of having a landscape strategy relates to sites of all sizes. Any large development site should have its own landscape network, while smaller ones should relate and create linkages within the wider strategy. So when setting out a new development site, some strategic landscape principles should be considered at the outset to create a more sustainable development. These include:

- Conduct a landscape assessment of the site, including a tree survey and a habitat survey. Identify sensitive areas and landscape elements that deserve retention, such as mature trees, natural watercourses, low-lying land (potentially good areas for natural drainage), unique habitats and ecosystems.
- Consider the local landform of the site. Avoid grading the site to a uniform slope for ease of construction and development but use it to create a site-specific response. This will preserve the natural ecology and hydrology of the site. When major grading occurs, the drainage pattern of the land is altered significantly and its downstream effects may be unclear.
- Reserve steep land that is otherwise unbuildable as escarpment and linear open spaces. These are unique landscape features that are part of the history of the site and are impossible to recreate at a later stage. They provide immediate amenity value to new residents and retain an existing feature within a newly emerging open-space network.
- Understand the local microclimatic conditions. What aspects can be exploited through design (i.e. south-facing slopes have a higher

solar-gain potential) and what aspects need careful attention to mitigate possible harsh conditions (i.e. buffer planting on the side of prevailing winds can minimise the reduction of the ambient temperature and heating requirements)?

Using these steps as a starting point for site layout will create a natural network of intrinsic landscape elements that give the site its own special character. They can become a focus in layout and design. Furthermore, they not only help create a development that responds to its setting and functions environmentally, but also a unique, site-specific solution that feels natural in its location. Once these elements are in place one can consider their detailed articulation to fulfil the strategic thinking behind them.

4.4   Drainage in city and landscape

A well-thought-out natural drainage strategy takes the points just raised further and has a profound effect on a strategic as well as a local level. We know that buildings, Tarmac, paving and parking seal vast areas of our cities. Little room is left for plants to take root. Statistics show that a startling 97 per cent of an inner urban city block is sealed and impermeable (see Table 4.1). Little to no water can trickle through to help replenish the groundwater. Instead, water is channelled into gullies and fed into the municipal sewers or straight out into our rivers, lakes and oceans. Reducing the amount of rainwater entering municipal sewers is positive as it means less wastewater needing treatment in sewage plants and hence less wasted energy (11).

Collecting surface-water run-off in gullies and flushing it away has become common practice nowadays. It is therefore not surprising that floods are more severe, have a shorter lead time and are more likely to occur more frequently. This again costs more energy and money in artificial prevention through flood defences.

To illustrate how the effect is compounded, consider this example: within the watershed (an area that drains into one system/river) of a small river, the developed/sealed land has increased threefold from 7 per cent to 20 per cent over 20 years (so it is still a suburban/rural form of development). The effect of the additional run-off has caused a fivefold increase in volume of water in the river and caused freak flood levels (13). Rather than allowing free drainage, water was collected in gullies and fed into the municipal system. More water has reached the rivers more quickly and this has been the main cause of flooding.

Rainwater is a relatively clean form of grey water (see Chapter 8). There is some contamination from the atmosphere and surfaces but much of this can be dealt with by fairly simple filtration systems using sand or soil. In many cases treatment will not need to be extensive. On-site natural drainage systems can help improve water quality and also reduce the risk of flooding.

Table 4.1
Degree of impermeability of different uses (12)

Inner urban areas	0.97
Dense residential areas	0.75–0.80
Mixed-use areas	0.80
Terraced housing	0.52
Semi-detached with small gardens	0.50
Semi-detached with medium gardens	0.42
Detached houses with large gardens	0.20

Many factors influence natural drainage, including:

- the amount of sealed and unsealed ground to be drained;
- the type of soil with its rate of permeability; and
- the intensity and duration of typical rainfalls and severe storms.

In the early stages of design it is important to allow for enough land to be able to implement a natural drainage system that can absorb all rainwater falling on all sealed and unsealed ground on the site. This can occur on a small scale on individual building plots, or larger development sites where a sophisticated interconnected water system can provide an important amenity function as well.

The possible retention systems are:

- low area drain;
- ditches and swales;
- permanent retention ponds; and;
- underground collection cisterns.

These are described in more detail in Table 4.2 and illustrated in Figure 4.6.

Depending on the specific aims to be achieved in the development, systems offer different advantages. The land-take necessary (the site area required for natural drainage systems) depends on the depth of the drain, which in turn depends on the local soil. The approximate land-take given in Table 4.2 is a percentage of the total site area to be drained and forms a rough guide. This enables the planning of natural drainage systems during the

Figure 4.6
Different retention systems

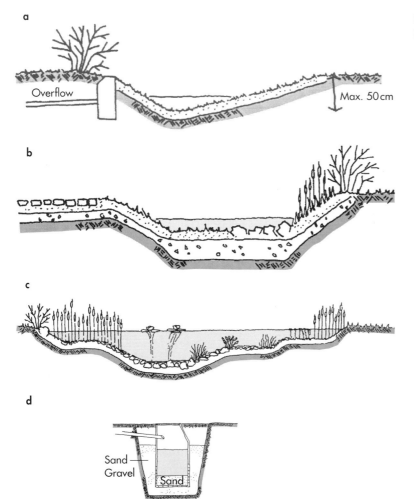

Table 4.2
Possible run-off retention systems to aid natural drainage

Retention system	General description	Approx. land-take	Max. depth	Construction cost	Environmental rating
Low area drain	Water collects at the lowest part of the site. It is generally designed as an extension of the lawn where the soil can become highly saturated during rain. After heavy rainfalls it can take up to three days for the water to drain away. Visually the area drain functions like a lawn and in summer can be used as such.	5–15%	0.5 m	Low	Medium/High
Ditches and swales	This can also just be grassed or planted with water-loving plants and laid out with rocks. The ecological value of ditches and swales is therefore higher since they provide a different habitat for animals. Ditches will form temporary water features after rainfall. They take up less space than a low area drain, as they can be a lot deeper – more than a metre. They are permanent features and cannot easily perform a different function when dry. Swales can form part of a surface water collection network that feeds into a pond.	2–5%	1.5 m	Low	High
Permanent retention ponds	These are permanent water features that have a higher amenity value than the above options. Ponds require more space than ditches as water levels need to fluctuate and more water is permanently retained on site. Retention ponds can be well supplemented with reed-bed filtration systems prior to discharge into a local river (see Chapter 9 for more detail). They score slightly higher than ditches and swales in environmental terms as they combine aesthetic values with new habitat creation and filtration.	3–5%	Varied	Medium	High
Underground collection cisterns	This is the most expensive option with the lowest environment rating and also limited capacity. Its strongest point is the minimal space required. Water is collected and slowly discharged into the groundwater through a sand and gravel pit. Filtration is therefore minimal and water quality inferior to that of the other systems.	>1%	N/A	High	Low

early stages of development. The exact land-take for a natural drainage system relates to the actual quantity of sealed land on site and the degree of impermeability of materials used (see Table 4.3).

Table 4.3
Degree of impermeability of different materials (14)

Pitched roof	1.00
Asphalt, concrete, paving with mortar joints	1.00
Flat gravel roofs	0.80
Paving on sand bed (tight joint)	0.80
Large paving with large joints	0.70
Mosaic paving with large joints	0.60
Bound gravel	0.50–0.40
Grasscrete paving	0.30
Planted roofs	0.30
Lawn	0.25
Vegetated area (general planting)	0.10–0.00

Further information on drainage is given in Chapter 9.

4.5   A case study: Granton Waterfront

To illustrate how all the principles discussed in this chapter can become part of the design solution for a new development, the Master Plan (15) for Edinburgh's new Waterfront at Granton forms an ideal case study. It is large enough in size to illustrate the strength of a landscape strategy, but also shows how sustainable drainage principles work in an urban context.

The site covers some 140 hectares of north-facing terraces overlooking the Firth of Forth. The 2 km coastline and its adjacent redundant industrial land is to become a new piece of the city with two local centres, ample employment opportunities, shopping, leisure facilities and education (see Figures 4.7 and 4.8). Within and around the two centres, more than 5,000 new homes will be created. In addition to this large portfolio of uses, some 30 hectares of open space will be integrated. This translates into a gross density of over 60 dwellings per hectare (dph), with variations from 30 dph to 150 dph depending on location (16). The plan incorporates retention systems based on low area drains, ditches and swales and permanent retention ponds. The first phases of this long-term project are just starting, but the principles and framework have been clearly set out in the Supplementary Planning Guidance that makes up the Master Plan.

The existing landform was a guiding design tool for setting out the framework for development (see Figure 4.9). The series of terraces, created partly naturally and partly through use, presented the opportunity to step the development up from the waterfront and maximise the splendid views over the Firth from within the development. The heavily planted break in level – the Brae – forms a natural continuation of the landscape buffer in the Green belt to the west and creates a natural corridor deep into the site. The land that formerly was the setting for Caroline Park (a wonderful historical relict) has been reserved to form the central focus and open space within the development. Further smaller local open spaces have also been introduced to link the southern parts of the site with the waterfront to the north. These "wedges" of open space link different natural and newly formed features of the Master Plan, creating an intricate network of open spaces of different character and ecological importance (see Figure 4.10).

Figure 4.7
Masterplan for Granton Waterfront

Figure 4.8
The diversity of uses at Granton
Waterfront

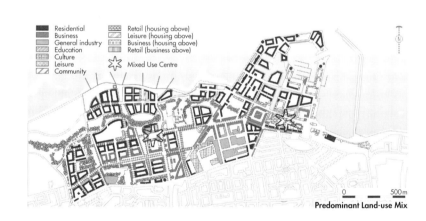

Figure 4.9
Terrain model of Granton Waterfront

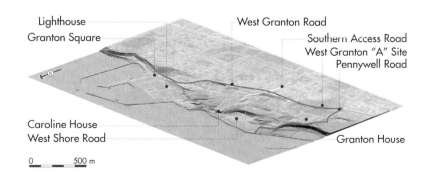

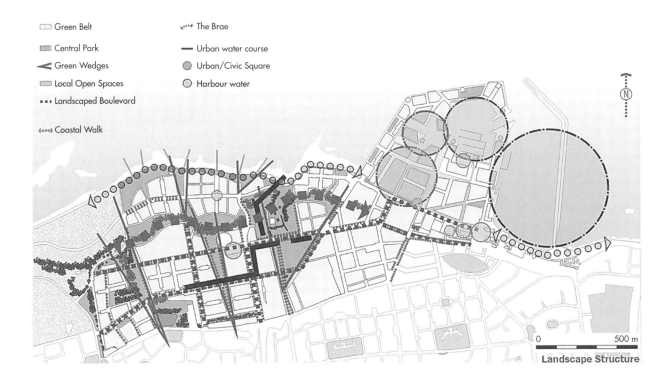

Landscape Structure

Investigations showed that a small creek, the Granton Burn, had been culverted in the past. Large areas of the site drain into it and discharge directly into the Firth. This untreated discharge from this part-industrial, part-abandoned land is largely responsible for the questionable water quality along this part of the shore. These findings illustrated the important task of the masterplan to improve the local water quality and bring back Granton Burn to form the basis of a sustainable drainage concept. All water joining Granton Burn will be naturally cleansed through plants and filtration and then discharged into the Firth. An additional filtration pond, which also has an important amenity function in the main park, and a swale along a local park stretching down towards the waterfront, further supplements this drainage system (see Figure 4.11). Because of the natural formation of the land, 70 per cent of the total site area will be drained through this system.

Figure 4.10
Landscape strategy for Granton Waterfront

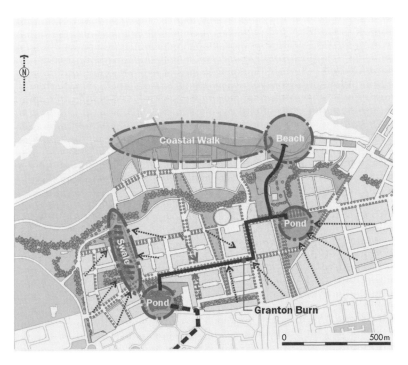

Figure 4.11
Sustainable drainage concept for Granton Waterfront

These strategic decisions made at the outset of the master planning process have set the framework for detailed solutions in the design and construction of part-areas of the plan. Following the same philosophy of creating a more sustainable development, detailed solutions will reinforce the strategic framework and enhance its functioning.

4.6 Conclusions

As this chapter has highlighted, landscape and vegetation have a profound role to play in our lives and our physical environment. It is therefore difficult to understand why these have been neglected in the development process over the past 50 years. New technologies and standardisation have driven the development process. Speed and cost considerations have required sites to be uniform and free of any obstructions. The effects of these practices have only become clear years after being adopted as "modern standards".

Only now are we noticing that many of the places we have created are too artificial with little regard to setting and context. Many people with the choice to move are leaving these places for more pleasant environments. We need to learn from the past and create developments with their very own sense of place, which take cues from their natural sites. The resulting development promises to be more humane and satisfying than "modern standards" have been. Considering the land, landform and landscape at the outset of site layout and design will set the scene for urban developments that fit into their settings. This major strategic decision will make numerous detailed options possible to lead to a more sustainable development overall.

4.7 Guidelines

1. Landscape and landform influence the appropriate form of development.
2. Landscape and vegetation influence the immediate microclimate by binding airborne particles, absorbing noise, raising humidity and minimising temperature fluctuations, as well as reducing wind speeds.
3. Trees and plants form a contrast to urban development and provide a calmer environment for urban life while views over green areas soothe the eyes and comfort the mind when tired. Parks, trees and planting can help to visually counteract the stress experienced in cities as a result of traffic, noise and pollution.
4. Creating wildlife corridors with interlinked habitats gives the best chance of attracting wildlife and improving the biodiversity of the city. A citywide landscape strategy can set the parameters for wildlife corridors to be developed and can make site-specific open spaces more meaningful.
5. The planning of any site needs to start with a landscape assessment, including topography and microclimatic conditions of the site. This will inform a unique development response with intrinsic landscape elements embedded in the place.
6. Reserve 5 per cent of a site to allow for free drainage or on-site water retention. Sustainable drainage best occurs on the lowest-lying land of the site.

REFERENCES

1. Champion, T. *et al.* (1998), Urban Exodus. CPRE, London.

2. www.futureforests.com

3. Fitch, J. M. and Bobenhausen, W. (1999), American Building: The Environmental Forces That Shape It. Oxford University Press, New York.

4. Lynch, K. (1971), Site Planning. MIT Press, Cambridge, Mass.

5. Hewitt, M. (2001), Can trees cut pain? Times, 4 September 2001, Section 2, p. 10.

6. Ibid.

7. Wauters, Amika (1999), Homeopathic Colour Remedies. Crossing Press, California.

8. Stadtentwicklungsbehörde (1997), Der Grüne Faden, Hamburg.

9. Llewelyn-Davies (2000), Urban Design Compendium. English Partnerships, London.

10. Stadtentwicklungsbehörde und Umweltbehörde (1993), Grünes Hamburg, Neue Ansätze und Strategien für eine ökologische Stadtentwicklung, Hansestadt Hamburg.

11. Edwards, B. and Hyett, P. (2001), Rough Guide To Sustainability. RIBA Publications, London.

12. Extracted from Haase, R. (1986) (see Reference 13).

13. Haase, R. (1986), Beiträge zur räumlichen Planung (14): Regenwasserversickerung in Wohngebieten: Flächenbedarf und Gestaltungsmöglichkeiten: Institut für Grünplanung und Gartenarchitektur, Hannover.

14. Extracted from ibid.

15. The Master Plan was drawn up by Llewelyn-Davies in association with Buchanan (CE) Limited, Parkman, Jones Lang LaSalle and Segal Quince Wicksteed.

16. Llewelyn-Davies (2000), Waterfront Granton Master Plan Volume II: Planning and Urban Design Guidance. The City of Edinburgh Council, Edinburgh.

5

Building design

Randall Thomas

5.1 Introduction

Our dreams and our desires are enacted in buildings – it is estimated that Europeans spend about 90 per cent of their time in them. As we have seen, almost half of our energy consumption takes place in them, hence part of their paramount importance for environmentally friendly design. Buildings, of course, also enhance the urban fabric/space. Without good architecture the "city" can splinter and become unable to sustain human well-being. Involving users and occupants in the design process will help to establish a feeling of participation and ownership (see, for example, Chapter 14).

Urban environments have a wide range of building types but two major categories – homes and offices – stand out and it is on these that we will concentrate.

Building design considerations in cities show how closely the buildings are linked to their surroundings. The temperatures, wind speeds, humidity levels, air quality and noise levels are all related to, and will depend on, density of development, energy sources, landscape, choice of transport systems and similar factors. Urban design and building design are inextricable.

5.2 An approach to design

Many architects would like to develop an environmental design that is architecturally satisfying and many are concerned with using the elements that allow the external environment to be modified to help them develop their architecture. We need good-quality buildings (and this can encompass a plethora of styles), better spaces between buildings and good sustainable design. Quality encourages people to care.

The environmental design of buildings is a blossoming field with many references available, some of which are listed below. In the future we can expect to see more regionalism in architecture as cities and buildings respond differently to their local climatic conditions. Northern temperate Europe is very different from the hotter, sunnier Mediterranean region and the hot, humid south of the USA is different from the colder clime of the northeast. The International Style with its ubiquitous (and sometimes beautiful) glass towers, heated and cooled with massive gulps of energy, has (probably) had its day. There is also a view that we should "touch the ground lightly" and minimise our environmental impact. One needs to ask – and answer – the question of the intended lifetime of the building and how it can flexibly adapt to the changes it will inevitably experience. One implication of this is that we are likely to see components that can easily be upgraded as the technology improves.

It also seems eminently reasonable to provide what people want: comfort, fresh air, natural light, views of the outside – the simple things in life, which contribute to our health (and, with a bit of luck, will increase our productivity).

Buildings should be designed from the inside out and the outside in simultaneously – designers need to imagine what the conditions will be for the occupants and the passers-by. Moving through the building should be an exciting, rewarding experience. To give but two examples, at the De Montfort Queens Building the occupants have often spoken of their enjoyment of the spaces – every walk through it is different (see Figure 5.1). And at the offices shown in Figure 5.2 the users have been delighted by a succession of proportionally related spaces with a variety of aspects.

Here we'll just touch on some of the key elements of building design. It is worthwhile considering the following:

A. Reduce demands

1. Reduce the demand for energy. This is often the most cost-effective thing to do and should always be the starting point. It is perhaps best viewed in terms of thermal energy (for space heating and domestic hot water) and electrical energy (for power and lighting).
2. Reduce the required energy for space heating by using the orientation, form and fenestration to make the most use of passive solar gain. At the same time the possibility of overheating needs to be reduced. This is discussed below. Lower the energy required by having a well-insulated envelope with high-performance glazing. Typically, U-values for walls should be in the range of, say, 0.19 to 0.26 W/m²K, i.e. better than current Building Regulations. Roof U-values could fall to, say, 0.10 to 0.15 W/m²K. High-performance glazing systems could achieve U-values of, say, 0.8 to 1.6 W/m²K. (And if you feel these figures are still too high, there is certainly design potential to improve on them.)
3. Seal the building tightly and ensure that it is pressure tested so that the air that enters does so by a known path. This will reduce the ventilation heat loss. The air that is provided should be of a high quality, which is a reminder that the urban air quality needs to improve and that materials within the building should not emit pollutants. Where reasonable, recover heat on the extract air (see Chapter 13 for one example).
4. Use less water, both hot and cold. Energy for domestic hot water can be reduced through judicious water use. This may involve spray taps and similar devices. Heat recovery on waste hot water is in the development stage and new products are becoming available.
5. Electricity is of course used for both "small" equipment, such as computers and photocopiers, and larger plants such as air-conditioning compressors and heating system pumps. As Max Fordham has said, "You really have to try hard to badly design a building for it to require air-conditioning"; none the less, many still succeed. How to reduce the need for cooling and how it might be provided are discussed below. Reducing other plant loads can be achieved by an inspired combination of architecture and engineering; "small" equipment loads can be reduced by the selection of everything from computers to refrigerators. If some cooling is still required, where practical, use an ambient source of coolth, e.g. borehole water or night-time air.
6. Daylighting (see below) can significantly reduce the demand for electrical energy.
7. Integrate the structure and the services so that energy use is minimised. Include thermal mass in order to even out both internal and external heat gains and to take advantage of night-cooling. What this means is that one can introduce cool night-time air, which might fall to, say, 13–15°C in the UK to counteract high daytime temperatures of, say, 24–26°C (see Appendix A). This can reduce peak internal temperatures the following day by 2–3°C. Note that

Figure 5.1
De Montfort Queens Building concourse

Figure 5.2
Entrance hall at the NFU Mutual and Avon Insurance Group Head Office

Figure 5.3
Thermal mass in the sinusoidal concrete
ceiling at the BRE Environmental
Building, Garston

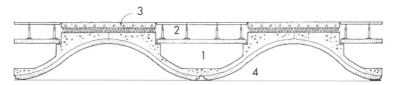

KEY   1. Ventilation paths through slab serving floor below
      2. Raised floor zone for cables and pipes for floor above
      3. Screed with heating/cooling pipes and insulation beneath
      4. Pre-cast concrete (75 mm) ceiling with *in situ* concrete topping
      5. High-level windows
      6. Side-hung casement windows
      7. Bottom-hung translucent windows

traditionally the mass has been integrated with the floor or ceiling
(see Figure 5.3) but it might be used in an innovative way in walls.
Disassociating thermal and structural mass may permit lightweight
superstructures.

8.   The embodied energy in materials should be reduced and waste on
       site should be minimised (see Chapter 7).

9.   The services with their controls should be energy-efficient and
       "intelligent" (but not necessarily more so than the people who will
       run the systems). Occupants should have some control over their
       environment. Boilers should produce extremely low levels of
       pollutants. Generally, all materials used in the services equipment
       should be environmentally friendly.

B. Provide energy in sustainable ways

Maximise the opportunities for capturing and producing energy. This
includes wind energy (see Chapter 6) but is principally directed towards
maximising the solar potential. Solar potential has four main aspects:

1.   Daylighting
       Daylighting and attractive views enhance the urban experience. In
       much of Europe, daylight needs to be better treated as the precious
       commodity that it is. To paraphrase Blake, "Light is eternal delight".
       The provision of natural light in buildings has been an ongoing
       concern of architecture. Lack of light and air, unsanitary and
       overcrowded conditions (excessively high densities in today's terms)
       and congested streets influenced urban planning from nineteenth-

century Paris to the Garden City movement (1). The benefits of daylight have traditionally included better health and a sense of well-being, as well as enlivening the architectural character of spaces. A varied, poetic mixture of daylight and artificial lighting is desirable. Describing the interior atmosphere of an English nineteenth-century home (Melsetter House by Lethaby), one critic praised it saying that it seemed "friendly not only to its occupants but to the very air" (2). Currently, providing more natural light in many spaces can reduce the CO_2 emissions associated with artificial lighting from conventional power sources. The ongoing improvement of glazing has also meant that it has been possible to provide more natural light with lower heat losses in winter. Glazing with a mid-pane U-value of $0.8\,W/m^2K$ (compared with single glazing at 5.6) is currently commercially available.

Successful spaces are those with good daylight – daylight from two sides, or a wall and roof, can be magical. The eighteenth-century Georgian home with its elegant proportions and tall windows facilitated gracious spaces. It served as a model for the London homes (built in the 1840s) shown in Figure 5.4 and is also of relevance to us today. Note too that the windows normally had wooden internal shutters, which were used at night to insulate to maintain comfort and to provide privacy.

Victorian schools used high ceilings and tall windows to bring light deep into classrooms and the principle still holds. In Germany, where there is an active effort to ensure office workers have access to daylight, the floor-to-ceiling height is about 3.5 m compared to a lowly 2.7 m in the UK (3). (The greater space can also be useful in providing flexibility for additional services if required during the life of the building.) Narrow floor plans will tend to ensure that sedentary workers are close to windows.

An intriguing modern approach to using a technical requirement (the need for larger glazed areas on lower floors to maintain the same daylight level because of obstructions from neighbouring buildings in urban contexts) as a generator for an aesthetic is shown in Figure 5.5.

2. Passive solar gain
Passive solar gain takes advantage of the solar radiation falling on roofs, walls and, particularly, windows. Letting in the sun during the heating season will help to reduce the demand for fuel. Conversely, in the summer the potential solar gain that could lead to overheating needs to be controlled at the façade (see, for example, Figure 5.6 where movable translucent louvres are used in the BRE Environmental Building) and by careful consideration of the ventilation path through the section.

What percentage of a façade should be glazed? There is no easy answer to this (architects will be relieved to hear). It will depend on the design intention, type of building, orientation, proportions of the interior spaces, U-value of the glazing and wall element, and the risk of overheating, to name just a few considerations. At the BRE Environmental Building, a careful analysis that included the value of daylight in reducing CO_2 emissions and seasonal heat loss through the wall and glazing led to a south façade with about 50 per cent window (5) (see Figure 6.9c). For modern housing, in part for reasons of privacy, a smaller percentage is common.

3. Active solar panels
These actively collect solar energy and transfer it to a fluid, usually water (see Chapter 6).

Figure 5.4
A London terrace

Figure 5.5
The Crystallography Building, Cambridge (4)

Figure 5.6
Movable louvres at the BRE Environmental Building

48

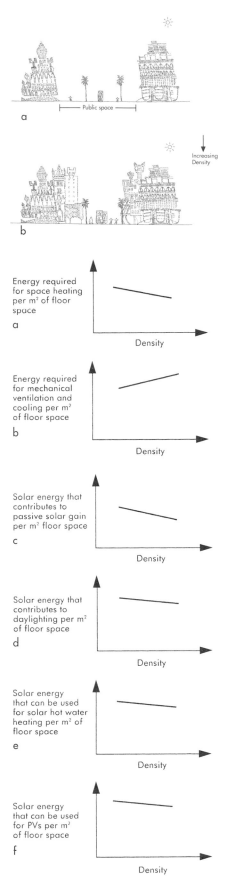

Energy required for space heating per m² of floor space

a

Density

Energy required for mechanical ventilation and cooling per m² of floor space

b

Density

Solar energy that contributes to passive solar gain per m² floor space

c

Density

Solar energy that contributes to daylighting per m² of floor space

d

Density

Solar energy that can be used for solar hot water heating per m² of floor space

e

Density

Solar energy that can be used for PVs per m² of floor space

f

Density

Figure 5.7
Energy and density considerations

4.   Photovoltaics (PVs)
      Developed in practical form especially by the US space programme, these devices convert solar energy directly into electricity (see Chapter 6).

Maximising the solar potential and dealing with ventilation in low-energy ways have important effects on form and we can expect to see more and more buildings that illustrate Mies van der Rohe's remark that "form is not the aim of our work, but only the result" (6). This will also apply to urban form because solar potential so depends on density of development, orientation, obstruction heights and other such considerations. Similarly, energy use will depend on some of these factors. Figure 5.7 (based on the simple, indeed, naïve, schematic of Chapter 1) sets out some very simple relationships between the main concerns as density increases.

If we first look at the factors that affect energy consumption, more compact developments will tend to have reduced heat losses (a); this is, for example, because there will be less surface area for the volume enclosed, there will be more shared wall space and because higher densities will encourage the heat-island effect. As density increases it will become more difficult to use natural ventilation solutions and for reasons similar to those given for heating the likely need for cooling is increased (b). More compact developments will be less able to use passive solar gain (c) and daylight (d) (in the summer, though, this will reduce the cooling load). We can see that generally, and it should be stated that the slopes of the lines in the figure are very notional, increasing density has a number of disadvantages.

In terms of suitability for solar thermal (e) and PVs (f), more compact forms will tend to mean that PVs can only be used on the roofs – less-compact forms will allow PVs to be used on façades. Solar thermal, in part because of plumbing considerations, tends to be used only on roofs. Roof heights are an important consideration. Where roof heights vary, and they usually do, putting taller buildings to the north is a way of preserving the solar potential. We can expect to see a sculpting of urban space as, for example, at Parkmount in Belfast (see Chapter 12).

Many of these issues have been studied quantitatively (7, 8) and an active group has produced a PV City guide (9). Very broadly, in a mixed grouping of homes and offices, all designed to a very high standard of energy efficiency and planned so that wherever possible the use of air-conditioning is avoided, more compact forms will tend to mean that somewhat more energy per unit of floor space is required. It will also be more difficult to supply this energy from the sun (as an extreme example, it is readily imagined that an isolated primitive hut, well-insulated and covered from head to toe with solar collectors and PV panels, could easily meet the occupants' energy demands from the sun). These are generalisations and need to be tested when looking at any particular development. It may be that providing all or most of the energy from the site itself is not actually that important if there are other sources of "green" energy available from outside the area (see Chapter 6). If it would seem that the situation is clear-cut and that higher densities result in a somewhat greater energy consumption in buildings, one should return to the overview provided by our systems approach of Chapter 1. It may also be that increased densities have advantages that outweigh their disadvantages. For example, it could be easier to use environmentally friendly community heating and CHP (combined heat and power) schemes. These may have lower distribution losses because of the higher densities and may be able to take advantage of economies-of-scale to extract energy from wastes (see Chapter 6). It is also likely to be easier to take advantage of large inter-seasonal energy storage. So that even if energy consumption appears to increase with density, it might still be worthwhile having a higher density because it would mean that energy could be supplied in a different, more efficient way, thus lowering the energy consumption.

As can be seen, the situation is rather complex and changes with time. And energy is only part of it. Privacy, the need for light, fresh air for health, land values, ease of rainwater recycling and waste collection, increased noise and many other factors will influence developments. In practice, in the future we are likely to see a whole range of solutions that will be more or less sustainable and in many ways this is very encouraging – we don't want to be strait-jacketed in developing our environments. The numbers will give the energy performance but of course that will only be part of the success story.

Building orientation and urban form are important issues because one wants, in a sense, to "think south". This will, of course, vary somewhat with building type. This structuring need not be too rigid and should certainly be balanced against other factors, such as the grain of existing streets and the effect on building appearance. But there will be a tendency to favour street patterns that run broadly east–west so that a building on the north side might point to, say, anywhere up to 30 or 40 degrees of due south and still have good solar potential (see Chapter 6 and Appendix A).

What can one say about the relationship between density, form and solar optimisation for housing? Traditional "two-up, two-down" nineteenth-century English workers' housing as shown in Figures 5.8 and 5.10(a) was built to a density of about 80–100 dwellings per hectare. (It has been suggested that densities up to 200 dwellings per hectare – corresponding to an "average obstruction angle" of about 30 degrees – are possible before negative energy impact is significant (10); "a" is an obstruction angle in Figure 5.8(a).)

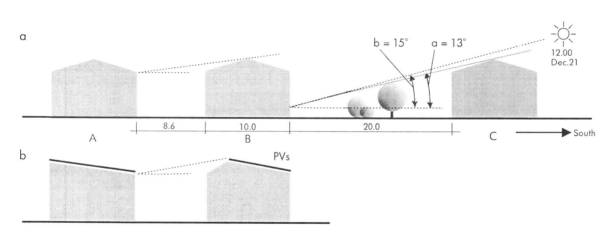

Figure 5.8
Density, form and solar optimisation
a. Traditional
b. Increased PV capacity

If the streets run E–W passive solar gain and daylighting are good on the rear floors of B. If active solar collection and PVs (see Chapter 6) are used on the roofs, the roof of A will be somewhat overshadowed by B during part of the winter but total PV potential can be improved by varying the architecture slightly, as shown in b.

From modest considerations such as these, and by keeping in mind other factors such as the existing streets (see, for example, Chapter 11), one could develop patterns.

Figure 5.9 shows courtyard housing, parallel slabs and a tower block, all built to the same nominal density of about 80–100 homes/ha; the floor area per home has been taken as 100 m². The courtyard housing can create a strong sense of place and community but will have some self-shading; interruptions in the form add variety and bring the sun in. The slab forms may feel rather soulless if not handled carefully architecturally, but can use the sun well. The tower block has a distinctive character with both advantages and disadvantages (see below); its solar potential (including the negative aspect of possibly overshadowing other buildings) depends very much on context. Orientation is shown as due south but considerable flexibility is possible (see, for example, photovoltaics in Chapter 6). Note also that these forms can be analysed in terms of the space the built form occupies on the site; the residual space can be examined to give, for example, an index of landscape potential. Overall, probably the most important development currently under way is thinking about form in terms of the total potential to use solar radiation – not only for passive solar gain and daylighting, as has been done traditionally, but also for active solar heating and electricity production through PVs. This is leading to a greater importance for roofscapes.

Figure 5.9
Urban forms

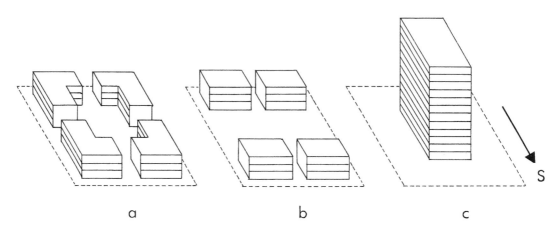

a   b   c

C. Allow for future adaptability to accommodate changes in use and advances in technology.

It is important that buildings can be altered within their lifetimes to allow for new technologies and social changes; indeed their lives will be extended if they can adapt. An example of incorporating technical developments is to plan for replacement windows that will lose less heat, admit more light and even produce electricity using photovoltaics. But this approach can be extended to roofs, internal partitions and external walls. This trend may lead to fewer buildings being conceived of as a work of fine art, much as, say, a Renaissance portrait by Raphael; instead we may see buildings that strive less to draw attention to themselves and contribute more to streetscapes and the comfort of their occupants while positively responding, environmentally, to evolving technologies. As one designer has noted, "One of the problems of modernism is that beautiful objects have been created that are not kind to people (11)".

5.3   Thermal comfort

Cities tend to experience higher temperatures in both winter and summer because of the heat island effect (referred to in Chapter 1). This is advantageous in the winter but potentially disadvantageous in summer, especially in offices where the likelihood of a need for cooling may be increased.

	Approximate Density Dwellings/ha	Comments
a	 80–100	Two-up, two-down homes (and shops) in Cambridge
b	 91	Cooper's Road (see Chapter 11)
c	 160	Camden Town, London

Figure 5.10
Urban densities

The occupants' ability to control their environment is known to be a key factor in the success of buildings. In homes this is usually achieved easily but in offices it requires careful attention both to design and to the mechanical heating, cooling and ventilation systems. Allowing temperature levels to vary tends to be both a more robust solution and a less energy-intensive one. Being able to adjust one's clothing is important in winter and summer and a strong case can be made that human beings enjoy the variability and stimulation. In the winter, temperatures between 20°C and 23.5°C are felt to be comfortable by many; in the summer, a comparable range is between 23°C and 26°C (12). It should be noted that the entire field of comfort is complex (humidity, ventilation and even noise can affect temperature preferences), subjective and controversial. In developing the brief for the BRE Environmental Building, another approach was taken, which was to provide the occupants with control over their environment (openable windows, movable shades, overrides on the building management system) and to limit the acceptable number of working hours when the temperature exceeded 28°C to 1 per cent of working hours and a similar figure of 5 per cent for 25°C (13). (In practice, the building performed better than the brief and 25°C was only exceeded 2 per cent of the time (14).)

5.4   Acoustic comfort

"Listen to the sound that fills the air" (15). The exceptional variety of sound and noise in cities varies from the exciting to the deplorable, the attractive to the deafening, the poetic rustle of leaves to the screech of braking buses.

In many homes, background noise levels, without the sounds of, say, stereos and televisions, will vary from about 25 to 35 dBA (Appendix D gives some basic data and some definitions). In open-plan offices, with conversations, telephones, and photocopiers, noise levels are in the range of 45 to 55 dBA. On busy city streets, with a mixture of cars, buses and the occasional motor bike, the noise level is likely to be 75 to 85 dBA at kerbside. Thus, attenuation of external noise is one of the most important issues for environmentally friendly buildings that rely on natural ventilation (see section 5.5). Of course, noise problems in cities are nothing new. One of the reasons Gilby House in Camden Town, London, designed in 1937 by Chermayeff, had a mechanical ventilation system was the noise from the traffic over cobblestone roads.

Another area of great importance is noise transmission – between buildings and, especially, between flats. Enjoyable urban living and acoustic privacy go together.

Noise from internal sources has historically been an issue with mechanical ventilation systems but a well-designed system, with acoustic attenuators if necessary, should perform well. Interior space planning and a judicious choice of office equipment should help control normal office noise.

5.5   Air and ventilation

Urban air quality has been and remains notoriously poor. In 1393, King Richard II empowered the University of Cambridge to improve certain gutters, which were "causing the air to be corrupted and many masters, scholars and others passing through the streets fell sick thereof" (16). Some current key contaminants are listed in Appendix C – they too have pernicious effects on our health.

What pragmatic designers need to consider is how they will bring air into the building. Natural ventilation systems have the advantage of requiring very little energy but may need more space for a low-resistance air path. A rough guideline is that ventilation from one side is adequate up to 6 to 8 m (and perhaps 10 m) and cross-ventilation is suitable for spaces 15–20 m across (17). In most housing, natural ventilation is the rule. Dual-aspect flats (that is, those laid out so that they have views to two sides) provide an opportunity for cross-ventilation and more access to the sun.

In offices and other buildings, though, the problems are often more complex. A range of solutions is being developed, running from assisted natural ventilation (or mixed-mode) approaches through to mechanical ventilation with assistance from the wind (18). A number of these methods incorporate heat recovery. It is also possible to add cooling to many of them with the source being a natural one such as groundwater (see Chapter 6). The use of the earth's cooling capacity is nothing new. Palladio describes the summer home in Costozza belonging to the Trentos, gentlemen of Vicenza. There cool air from underground excavations from former quarries was led into the building by a system of underground passageways (19).

Ventilation strategies

Figure 5.11 shows a variety of ventilation strategies.

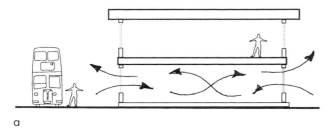

a

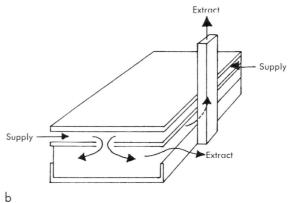

b

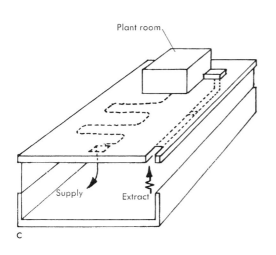

c

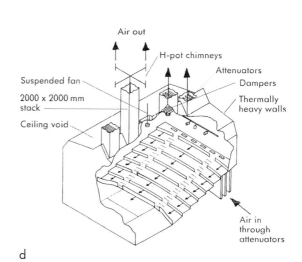

d

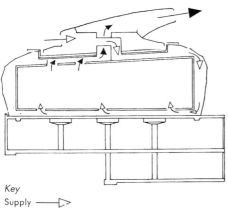

Key
Supply ——▷
Extract ——▶

e

f

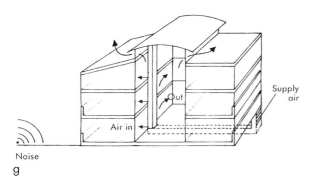

Noise

g

Figure 5.11
Selected urban ventilation strategies
a. Air in and out from perimeter
b. High-level supply; extract at mid level or high level,
   e.g. stacks (see also Figure 7.3)
c. Mechanical ventilation (i.e. supply air path
   incorporated in thermal-mass deck)
d. Air in from perimeter, extracted via stacks (Contact
   Theatre)
e. Wind-assisted high-level supply and extract
f. Ventilation via a quiet courtyard
g. Ventilation from a protected area

Figure 5.12
Cartier Foundation

Figure 5.13
A North American park

Figure 5.14
Attenuators and student at the Contact
Theatre, Manchester

Many of the methods shown are "sustainable". However, one needs to examine the comfort achieved, the energy required and the corresponding CO_2 emissions. Hybrid solutions combining natural and mechanical ventilation might be a way of dealing with variable urban pollution rates (20); a mechanical boost might be provided when pollution was high and filtration, say, was needed. In a similar way, intelligent controls could shift the ventilation mode according to noise level if required. These more sophisticated approaches could function on plan, reacting to local roads, say, or vertically in tall buildings.

Jean Nouvel's brilliant gambit at the Cartier Foundation in Paris (Figure 5.12) was to set a glass (unfortunately air-conditioned) building back behind a glass screen. The building is thus protected by its distance (probably more than the rather perforate screen) from much of the noise and pollution of the busy Boulevard Raspail. Perhaps what is most important though is the integration and richness of the combination of architecture and nature because the screen allows one to see, in Nouvel's words, not only a tree but the reflection of the tree (21). (And, indeed, the same is true of buildings, pedestrians and parked cars.) However, this strategy is not often an option.

The multiplicity of effect of the Cartier screen is also seen more naturally when water is brought into the city, as shown in Figure 5.13.

A more common urban ventilation strategy is the traditional one of a building close to the road with air introduced at the perimeter – this is satisfactory provided that noise and pollution levels either are not excessive or are dealt with in some way (such as with attenuators for noise). Figures 5.11a to 5.11e show several options (among many available) with air introduced and exhausted in a variety of ways (see also Figure 5.15 and Chapter 13). Figures 5.11f and 5.11g show other strategies, with air brought in from courtyards or quiet side areas. The appropriate option will depend on the context, as always.

The ventilation strategy at the Contact Theatre, Manchester (22, 23) (see Chapter 15), incorporated a low-pressure drop inlet with large acoustic splitters (that is, attenuators) to reduce the urban noise level to an acceptable level for stage productions. Figure 5.14 shows a visiting mature French student of environmental design experiencing the sound reduction at first hand.

Façade and section

Note that the façade and section need to work together. Passive solar gain, overheating, ventilation, views, thermal mass, acoustics, aesthetics and many other factors meet here, not necessarily always happily.

If we take a south-facing façade it will need to have a way of controlling solar gain. This may be through external movable louvres (see Figure 5.6) but for many buildings it is more likely to be through, say, retractable internal blinds or shades. Where the budget is more substantial and where life-cycle costing is encouraged, these blinds may be in between two layers of glazing.

If air is to be brought in from the façade, solar control measures should not interfere with the air path. One solution, where a raised floor is needed to distribute services, is to use the void for air distribution. This was done at Granta Park, shown schematically in Figure 5.15, which is, in fact, in a suburban business park, but the principle can be applied to more urban spaces. If noise is an issue, attenuation can be placed in the void; if filtration is required, a more powerful fan could be used to overcome the additional resistance. The air enters via floor grilles. After passing through

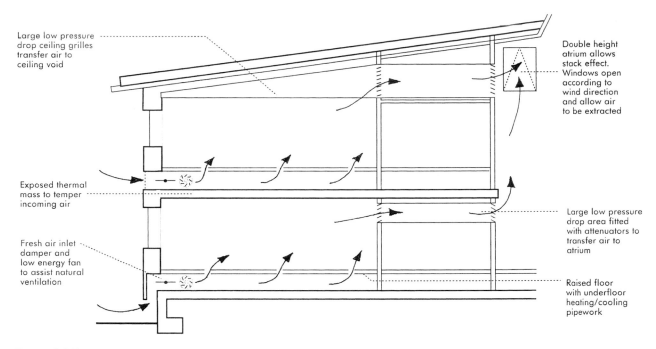

Large low pressure drop ceiling grilles transfer air to ceiling void

Double height atrium allows stack effect. Windows open according to wind direction and allow air to be extracted

Exposed thermal mass to temper incoming air

Large low pressure drop area fitted with attenuators to transfer air to atrium

Fresh air inlet damper and low energy fan to assist natural ventilation

Raised floor with underfloor heating/cooling pipework

Figure 5.15
Schematic section showing the ventilation path and thermal mass at the McClintock Building, Granta Park

the space, the air rises into the ceiling and then out into an atrium, from which it is exhausted. In this case the floor-to-ceiling height was about 2.8 m. Higher ceilings will allow greater penetration of daylight. The thermal mass in this design is in the floor void and to a lesser extent the walls; on the ground floor there is also thermal mass in the ceiling.

5.6 Roofs

One approach is to have a "green" roof, which consists of vegetation – this can be a thin layer of, say, 50 mm with a light loading of about 60 kg/m^2, which structural engineers will appreciate. The plants can include sedums, herbs and grasses. This was done at the Almeida Theatre at King's Cross, London (see Figure 5.16 and Chapter 14) for both visual reasons and acoustic ones – the soil gave just the right degree of attenuation required for performances. Thermal performance of the roof was also improved somewhat.

The advantages of soil and plants on a roof include the retention of rainfall during storms, thus reducing the risk of flooding. Plants, as we have seen, will also take up CO_2, produce O_2, release moisture into the atmosphere, remove dust particles, provide an environment for wildlife including birds and butterflies, gladden our lives and remind us of the seasons – all in all a bargain at 60 kg/m^2.

An alternative is to have a roof that can take a high loading (say, concrete) and landscape the roof with, say, 300–500 mm of topsoil and an extensive variety of plants and shrubs, effectively a traditional roof garden.

But of course the green roof may compete for the same area as a solar collector – for daylighting through rooflighting, for passive solar gain, for active solar hot-water heaters and as a photovoltaic panel (see Figure 6.6). Which strategy is appropriate – or, perhaps better, which mix, will depend on the project.

Figure 5.16
The green roof at the Almeida Theatre, King's Cross

Figure 5.17
Roof in Munich

We may also see some unusual roof shapes such as this (non-solar) roof in Munich (Figure 5.17), as buildings are likely to attempt to emulate (on a modest scale) the heliotropic movements of plants towards the south to increase their solar gain.

5.7   High-rise buildings

Le Corbusier drew high-rise buildings set in an urban park but his vision of the city was a complex one and he said that "family life would hardly be at home [in a city of towers] with their prodigious mechanism of lifts" (24).

A key question when considering tall buildings is to ask whether they are being proposed out of desire or necessity. And if desire, whose desire? The images of strength and virility associated with tall buildings (and perhaps now challenged by the attacks on the World Trade Centre) may suit both politicians and architects but their role in a sustainable community requires careful analysis. Perhaps the way forward is for mixed high- and low-rise developments, as at Coopers Road (Chapter 11) and Parkmount (Chapter 12). A mixture of uses (homes, offices, shops, leisure) to create semi-autonomous communities may also make "high rise" more sustainable. What is very clear as one studies London is that tower blocks only work both socially and environmentally if the space between them is successful and if they are well managed.

Space limits us to discussing only a few of the environmental considerations associated with them.

Solar considerations

Light is an important issue. If a tower block deprives the adjacent buildings of light and thus of the opportunity to use photovoltaics and active solar thermal, problems might ensue. In energy terms, the PVs that might be mounted on the south façade of the tower, however, might more than compensate for the energy loss of any buildings overshadowed by the tower. This would, of course, require some analysis.

Solar considerations may lead us to putting higher buildings at the north of a site (provided, one would hope, that they don't overshadow their neighbours) unless they fulfil another function such as marking particular sites in a broader urban context.

As the use of solar energy becomes increasingly viable (and necessary), we can expect to see legislation to control urban forms to enhance solar access in a similar way that daylight access has influenced cities in the past. For example, much of Paris consists of areas of great beauty, planned and built in the nineteenth century with light as an important generator of form.

Wind considerations

Tall buildings may experience high wind speeds at both the top and the bottom. At the top this is a potential source of energy (see Chapter 6). However, if, say, winds hit the building and cascade down, this is likely to annoy pedestrians. Modelling, either in wind tunnels or by CFD (computational fluid dynamics), can help at the early stage of analysis to avoid problems. Design and planning can assist to mitigate negative effects. The Success House tower (see Chapter 11) is set on a plinth to protect pedestrians at street level.

Energy consumption

Tall buildings will tend to have a greater surface to volume ratio and hence an increased heat loss. Exposure at higher levels is also likely to increase the space-heating energy use. Lifts will of course require energy.

A detailed study of the various options for both energy use and energy supply will need to be carried out for any particular site. And we need to keep in mind that if we increase density with high-rise buildings we are likely to reduce the energy associated with transportation.

5.8   Plants in buildings

Plants in buildings will require water and nutrients but have similar advantages to those outside: CO_2 uptake, oxygen production, a humidifying effect and the psychological effects of their presence and colour. In addition, plants can remove a number of indoor air pollutants, for example, bamboo palms will remove formaldehyde; lady palms, ammonia; and areca palms, toluene (25). Other plants will remove nitrogen dioxide, carbon monoxide and benzene.

5.9   Guidelines

1.   Always do the things that cost "nothing" first, such as getting the orientation, form and massing correct.
2.   The form of the building will be the result of considering many factors, from street pattern through to energy consumption and the potential for energy production.
3.   Buildings should be well insulated and tightly sealed. The demand for energy should be reduced.
4.   Design for daylight.
5.   Occupant satisfaction is an important element of success. Buildings and cities are for people.
6.   The use of low-energy ventilation (and, in some cases, cooling) systems is vital.
7.   The building's main components, including façades and roof, should be conceived of and constructed so that they can be altered over time to suit new uses and new technologies.
8.   There are many ways to design sustainable urban communities. Success will come with the right balance (or something entirely opposite and extraordinary).
9.   Design buildings so that they possess "the lightness and joyousness of springtime which never lets anyone suspect the labours it cost" (Matisse) (26).

REFERENCES

1.   Vidler, A. (2001), A City Transformed: Designing "Defensible Space". The New York Times, 23 September, p. 6.

2.   Garnham, T. (1993), Melsetter House. Phaidon, London.

3.   Anon. (2001), The heights of design. Light and Lighting, Issue 25, p. 3.

4.   For further information on this building see Ray, N. (1992), Crystal Clear. Architects' Journal, 8 July, pp. 24–34.

5. Thomas, R. (1999), Environmental Design. Spon, London.

6. Cited in Schulze, F. (1985), Mies Van der Rowe. University of Chicago Press, London.

7. Steemers, K. (2000), PRECis: Assessing the Potential for Renewable Energy in Cities. University of Cambridge: the Martin Centre. The European Commission Joule III (DGX11) Contract JOR3 – CT97–0192.

8. Steemers, K. (2000), The paradox of the compact city. Architects' Journal, 2 November, pp. 42–43.

9. Barker, M. et al. (2001), Solar ElectriCity Guide. Energie, European Comission; Institut Cerda, Barcelona.

10. As reference 8.

11. Professor Klas Tham speaking at the RIBA Conference Sustainability at the Cutting Edge, 22 October 2001.

12. Berglund, L. (1998), Comfort and humidity. ASHRAE Journal, August, pp. 35–41.

13. Anon. (1995), A Performance Specification for the Energy Efficient Office of the Future. BRECSU, BRE, Garston.

14. Anon. (1999), Garston Revisited. Natural Ventilation News, September, p. 2.

15. Logue, C. (1994), War Music. Faber, London.

16. Petty, M. (1995), There was something in the air. Cambridge Weekly News, 20 September, p. 6.

17. As reference 5.

18. Richardson, V. (1998), Air Force One. RIBA Journal, September, pp. 68–71.

19. Palladio, A. (1997), The Four Books on Architecture. The MIT Press, London.

20. Kukadia, V. (2001), External influences on indoor air quality. Building Services and Environmental Engineer, November, p. 19.

21. Jean Nouvel speaking on French radio (France Inter) on 21 December 2001.

22. Palmer, J. (1999), First Contact. Building Services, 21 (10), pp. 31–34.

23. Garnham, T. (1999), Building – Alan Short in Manchester. Architecture Today; No. 99, pp. 24–39.

24. Windsor, P. (1977), How wrong was Corbusier? The Architect, April, pp. 36–37.

25. Hazell, T. (1997), The green protectors. Daily Mail, 23 January, p. 61, citing research in Wolverton, B. C. (1997), Eco-Friendly House Plants, Weidenfeld and Nicolson.

26. Quoted in Cork, R. (2002), "A Master for the Lord". The Times, 2 January, p. 15.

FURTHER READING

Air input from high level: Hatton, B. (1999), Building – Awopbop-aloobop-wopbamboom: an icon for Sheffield. Architecture Today, 97, pp. 24–30.

Air path incorporated into pre-fabricated structural elements: Sensitive Addition to a Campus. Architects' Journal, 15 June 1995, pp. 31–38.

Tall buildings: www.skyscraper-conference.de. From the Trends in Tall Buildings Conference, Frankfurt, Germany, 5–7 December 2001.

Anon. (2001), Better Places to Live: By Design. DTLR and CABE, London.

Anon. (1989), The Assessment of Wind Loads. BRE Digest, p. 346.

Bordass, B. and Jaunzens, D. (1996), Mixed Mode – The Ultimate Option? Building Services, November, pp. 27–29.

Carl, P. (2000), Urban density and block metabolism. In Architecture, City, Environment, Proceedings of PLEA 2000, Cambridge, UK. James & James, London.

Eaton, K. and Ogden, R. (1995), Thermal and structural mass. Architects' Journal, 24 August, p. 43.

Edwards, B. and Turrent, D. (2000), Sustainable Housing. E&FN Spon, London.

Hall, P. (1999), Sustainable cities or town cramming? RSA Journal, 4/4, pp. 73–81.

Littlefair, P. J. et al. (2000), Environmental site layout planning: solar access, microclimate and passive cooling in urban areas. BRE, Garston.

Lloyd Jones, D. (1998), Architecture and the Environment, Bioclimatic Building Design. Laurence King Publishing, London.

Ogden, R. and Kendrick, C. (1998), Using fabric thermal storage to provide passive cooling. Building Services, May, pp. 47–48.

Pidwill, S. (2002), Learning with Louvres. Architecture Today, 125, pp. 26–34.

Steemers, K. (2001), Urban Form and Building Energy. In Echenique, M. and Saint, A. (Eds.), Cities for the New Millennium. Spon, London.

Various authors (2002), Tall Buildings. Building Services, 24 (3), pp. 40–57.

Yeang, K. (1999), The Green Skyscraper. Prestel, London.

6

Energy and information

Randall Thomas

6.1   Introduction

The energy situation is volatile and not easily predictable. However, if one doesn't believe that nuclear energy is our future, one is essentially left with the transition to a solar society (solar is used here in the broad sense of renewable energy sources, including wind and biomass, which are effectively derived from solar energy). As we saw in Chapter 1, at present renewables supply only about 1.5 per cent of our energy. In the UK the intention is to increase this and, as an example, the government is aiming to supply 10 per cent of electricity from renewables by 2010, mainly through an increase in wind power.

What form a solar society might take and what the time-scale is remain to be seen. "A solar future in the near future" is a rallying cry for many. The principal options for forms are shown in Figure 6.1.

In Figure 6.1a, solar generating stations using renewable energy sources such as photovoltaics (PVs) or wind turbines generate electricity that is then distributed via a grid system to users. Figure 6.1b shows a similar arrangement but here the electricity is used to split water into hydrogen and oxygen. The hydrogen is then distributed via a pipe network as natural gas is now. This approach is sometimes referred to as "the hydrogen economy" (see below).

Figure 6.1c shows solar energy being produced within the urban area itself. This has the advantage of producing energy near the point of use and so reducing transmission losses. However, it has a number of difficulties, some of which have already been referred to and others of which will be discussed below.

Figure 6.1d shows a mixture of "on-site" generation, with the remaining demand being supplied by a grid system. This has a number of advantages, including being the most pragmatic and the most adaptable, and so is the one we are most likely to see developed in the short term. In the longer term, we are likely to have a hydrogen economy in which electricity and hydrogen both play a role.

One of the encouraging aspects of sustainable design is the diversity that will result naturally and so in England a supplementary source of electricity will be chicken litter while in Thailand there are plans to produce biogas from agricultural waste water and elephant dung.

Supply and demand

How much energy is available? In the London area, approximately 950 kWh are incident on every square metre of horizontal surface each year. The energy in the wind (available to a typical wind turbine) is very similar at approximately 830 kWh per annum per square metre (of vertical "surface"). (Appendix A gives incident radiation data for the British Isles and Europe and Appendix B gives wind data for the British Isles.)

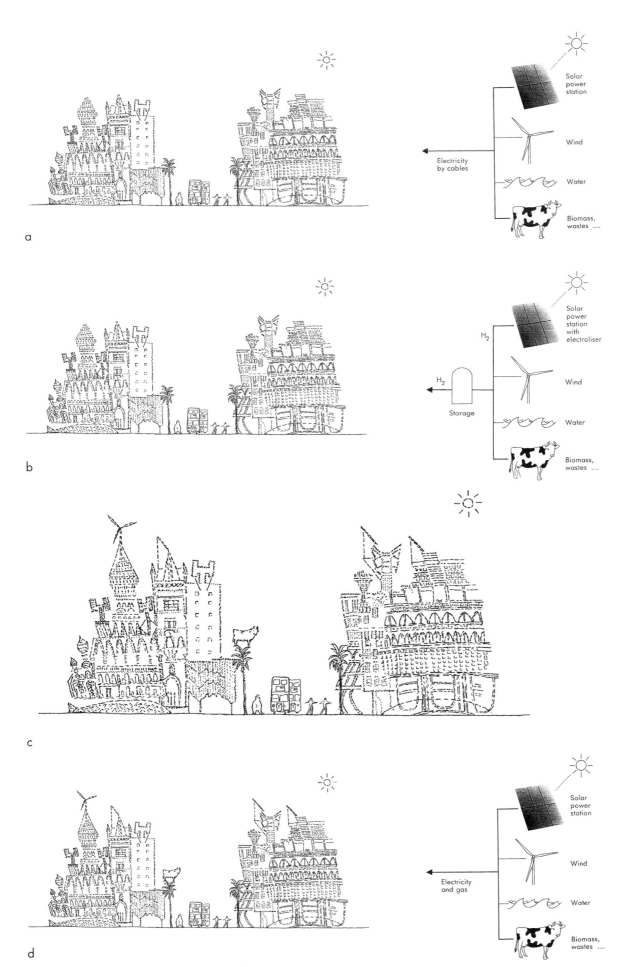

a

b

c

d

Figure 6.1
Solar options

Figure 6.2
Delivered energy use and CO_2
production at Coopers Road
a. Coopers Road
 Delivered energy use (i, ii)
 100% = 116 kWh/m²y
b. Coopers Road
 CO_2 production
 100% = 31 kg CO_2/m²y
c. BRE Environmental Building
 Delivered energy use
 100% = 83 kWh/m²y (iii)
d. BRE "Improved Version"
 Delivered energy use
 100% = 42 kWh/m²y (iv)

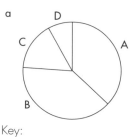

a

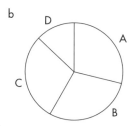

b

Key:
A. Heating 37% (43 kWh/m²y)
 • Fabric heat loss
 (25 kWh/m²y)
 • Ventilation heat loss
 (18 kWh/m²y)
B. Domestic hot water (39%)
 (45 kWh/m²y)
C. Power (16%) (19 kWh/m²y)
D. Lighting (8%) (9 kWh/m²y)

Key:
A. Heating 29% (9 kg CO_2/m²y)
 • Fabric heat loss
 (5 kg CO_2/m²y)
 • Ventilation heat loss
 (4 kg CO_2/m²y)
B. Domestic hot water (29%)
 (9 kg CO_2/m²y)
C. Power (29%) (9 kg CO_2/m²y)
D. Lighting (13%) (4 kg CO_2/m²y)

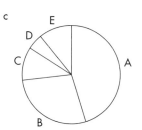

c

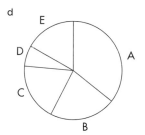

d

Key:
A. Heating 45% (38 kWh/m²y)
 • Fabric heat loss
 (25 kWh/m²y)
 • Ventilation heat loss
 (13 kWh/m²y)
B. Small power 28%
 (23 kWh/m²y)
C. Lighting 11% (9 kWh/m²y)
D. Service power 5% (4 kWh/m²y)
E. Hot water 11% (9 kWh/m²y)

Key:
A. Heating 36% (15 kWh/m²y)
 • Fabric heat loss
 (8 kWh/m²y)
 • Ventilation heat loss
 (7 kWh/m²y)
B. Small power 22%
 (9 kWh/m²y)
C. Lighting 19% (8 kWh/m²y)
D. Service power 7% (3 kWh/m²y)
E. Hot water 16% (7 kWh/m²y)

Notes for Figure 6.2

i. Largely based on data kindly provided by ECD – Energy and Environment.

ii. Note that the distinction between Figure 1.6 and this figure is one of primary energy and delivered energy. For converting delivered energy to useful energy (see, for example, Figure 6.18) we assume all of the delivered electrical energy is useful and that boiler efficiencies are 85 per cent. Power includes some cooking.

iii. Based essentially on a performance specification for the Energy Office of the Future Report 30, BRESCSU, Garston, Watford. Note that for comparison CO_2 production figures have been updated to be consistent with figures of 0.47 kg CO_2/kWh of electricity and 0.20 kg CO_2/kWh of gas.

iv. Estimates by Max Fordham LLP of feasible reductions. Ventilation heat loss assumes heat recovery on extract air.

How much energy is needed? This, of course, is a much more difficult issue. Many buildings in urban areas are used for either offices or housing. Figures 6.2a and 6.2b show the energy demand and CO_2 production, respectively, of the low-energy housing planned for Coopers Road in the London Borough of Southwark (see Chapter 11). (These figures are for a "base case" model; an "improved" version has an estimated energy demand of $93\,kWh/m^2y$. Note that it is interesting to compare this with Malmö (Chapter 17). Figure 6.2c shows the energy demands of the low-energy BRE Environmental Building, which although outside London is easily adapted to an urban environment; Figure 6.2d is a hypothetical "improved" version.

To put this in perspective, each home at Coopers Road will produce about $2,500\,kg$ of CO_2 a year. A pair of homes will annually be close to the weight of an elephant (about $5,000\,kg$). We really want to be more like butterflies on our earth (Figure 6.3) and a strategy is discussed in Chapter 11 for achieving this.

Generally, for housing it should be possible to reduce the space heating demand to, say, $10–15\,kWh/m^2y$ but there will always be some need for energy. There will be some fabric heat loss and a ventilation heat loss that can not be eliminated since even if heat is recovered the process will be somewhat inefficient. Hot-water requirements might be reduced to $300–400\,kWh/y$ per person but there will be a demand and one has the same issue with heat recovery. With regard to electricity we should be able to reduce the need to, say, $10\,kWh/m^2y$ by using low-energy appliances and by greater use of daylight.

It is interesting to note also the sensitivity of demand to urban density and some indication of this is given in Table 6.1.

Thus, very approximately, two-thirds of the energy demand in housing (see Figure 6.2a) is independent of density and only one-third will be affected.

We can now examine in more detail an energy strategy for an urban site. In the background, as always, is the concept of reducing demand and then meeting that demand in a suitable way. Our first appropriate solution will be the use of solar energy.

Figure 6.3
Butterflies and elephants (not to scale)

Table 6.1
Energy demand for housing as related to density

Item	Comments
Space-heating load/person	
a. Fabric heat loss	Decreases with density
b. Ventilation heat loss	Weak connection but will be reduced slightly by urban heat-island effect
Domestic hot-water load/person	Independent of density
Electricity for lighting	Will increase somewhat with increasing density
Electricity for small power (appliances, computers, etc.)	Independent of density
Electricity for lifts	Will increase with density

6.2   Solar energy

In Chapter 5 we looked at maximising the solar potential. Here we will look at active solar thermal and photovoltaics in more detail.

Solar thermal

With solar thermal, the sun's energy is actively collected by either water or air and the heat is then used inside the building.

Solar water heating has been with us for well over a century. Early photographs of the urban roofscape in Los Angeles, California, in 1900 show solar collectors fitted on roofs (see Figure 6.4) (1).

Figure 6.5 shows a more recent solar-thermal array for hot water to the kitchen, installed in the 1970s on the roof top of a hotel in Cambridge, England.

Figure 6.6 shows the Zero Energy House in Amersfoort and is indicative of how many of our future roofs will be used for solar thermal, PVs and daylighting.

Small-scale solar-thermal systems in the London area can be expected to provide about 400 kWh/m²y. Thus, if 4 m² of panel were installed on the roof of an average home at Coopers Road (with 4.23 people), it would provide a significant contribution (47 per cent) to the hot-water demand. As output falls off in the winter, the backup would be provided by the main heating system.

Economies of scale are possible and a number of countries have used large thermal stores as shown in Figure 6.7.

The potential for solar thermal is significant and it will undoubtedly be one of a number of ways of moving towards more sustainable cities.

Photovoltaics

Photovoltaics (PVs) are environmentally friendly systems that produce electricity directly from solar radiation. The PV phenomenon was discovered in 1839 by Antoine Becquerel but it is in the last 50 years or so that extraordinary progress has been made, driven by research and development in the space and computer industries. PVs are a very low-polluting, established and reliable technology and are found everywhere from marine buoys to solar planes (see Figure 6.8). Appendix A provides a slightly more technical introduction.

In buildings, PVs are in use on roofs and walls (see Figure 6.9) and can also be parts of sunspaces (see Chapter 16) and sunshades. The optimum orientation and tilt in the UK is due south and at an angle from the horizontal of about the latitude minus 20 degrees. So in the London area the optimum tilt to maximise the year-round production of electricity is about 30 degrees. However, there is considerable scope for flexibility and orientations within about 30 degrees east or west of due south and tilts of 10 to 45 degrees will give approximately 95 per cent of optimum performance (see Appendix A). This can be important in the layout of urban streets and in optimising the solar potential of the site. Generally, one should try to avoid overshadowing of the PVs by other buildings (and parts of the building itself) and obstructions including trees. This will argue for putting PVs on roofs unless the southerly facing walls are relatively unobscured. The Parkmount development in Belfast (see Chapter 12) is an example of how to model the urban space to use solar energy for photovoltaics. It is also an example of how one can quantify the extent to which the PV potential of the site has been realised (see Appendix A).

The most promising PV systems are what is known as grid-connected ones. These are currently more cost-effective because they supply excess energy to the grid rather than store it in, say, batteries.

a

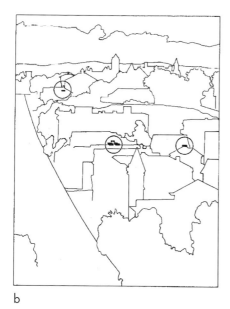

b

c

Figure 6.4
Solar panels in Los Angeles, California, 1900

a. View from a distance
b. Positioning sketch
c. Detail

a

b

Figure 6.5
Solar-thermal array in Cambridge, England
a. View from a distance
b. Detail

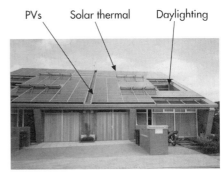

Figure 6.6
Roof of the Zero Energy House,
Amersfoort

In the future we can expect to see a wide range of products, in varying module sizes (from roof-tile dimensions to, say, $2 \times 4\,m$ plates) or varying colours, for use on façades and roofs. A number of manufacturers are developing "hybrid" PV and thermal-solar panels that recover the heat from the back of the PVs; such a system is being installed at the facilities for Renewable Energy Systems at King's Langley (3). It is also likely that windows incorporating PVs that transmit visible radiation but capture the infra-red (see Glossary) will be available.

Certain forms such as railway stations may also be particularly suited to PVs because of their large roof expanses in often relatively open urban space.

One area that will become extremely important is the solar potential equivalent of rights-of-light. Clearly, one will not want to invest in PVs and solar thermal only to see the output reduced by future developments if they overshadow the solar installations.

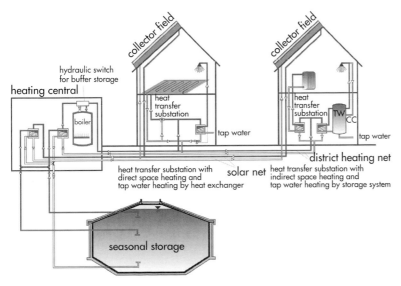

Figure 6.7
Large-scale thermal storage at Friedrichschafen, Germany (2)

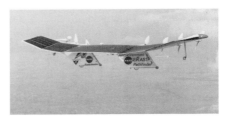

Figure 6.8
Solar plane

Figure 6.9
PVs on buildings

a. The Netherlands National Environmental
Education Centre

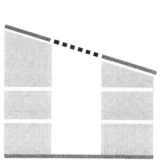

Atrium

b. Doxford Solar Office

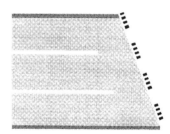

Inclined wall with windows

c. The BRE Environmental Building

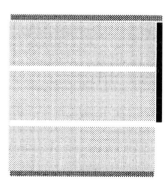

Vertical

There are a number of different types of PVs but high-quality monocrystalline with an efficiency of 12–15 per cent in a grid-connected system will provide very roughly 100 kWh/m²y when installed facing due south and at a tilt of 30 degrees.

Costs

Almost everyone agrees that PVs will become less and less expensive and that they will become economic – the only matter of debate is when. PV manufacturing capacity is expanding rapidly and this should help reduce costs.

At present, electricity from small to medium PV systems costs about 25–35 p/kWh. Some predictions indicate a cost in 2020 of approximately 10–16 p/kWh (see Table 6.5).

6.3   Wind energy

At a time when large wind turbines are being constructed around Europe and the price of energy from them is becoming more and more competitive, one may ask why bother considering smaller urban versions.

One reason is that although wind farms can be things of beauty, some consider that they spoil the countryside visually and acoustically and that Wales, for example, would be better off without them. For off-shore wind farms there are the inevitable issues of construction and maintenance costs as well as the need to transmit the energy produced to users who are likely to be hundreds (if not more) of kilometres away. However, wind generation is now commercially viable in many parts of the world, including numerous areas in the UK.

Of course, in urban environments there are noise and safety issues that need to be considered. There is, none the less, wind in the city. How much is there? And how much energy is available?

In Appendix B there is a map of the availability of wind energy in the UK. In the London area the mean annual wind speed is approximately 4 m/s at a height of 10 m. The effect of the urban environment on this is complex. Generally, there will be a reduction of wind speed due to the uneven urban terrain. However, this may be compensated for by an increased wind turbulence in some parts of cities (other parts, such as urban courtyards, will be sheltered – and deliberately so). Buildings also may serve as masts for wind turbines where height and turbulence may be advantages. Figure 6.10 shows a simplified diagram of wind speeds.

Small-scale wind-turbine systems have overall efficiencies of approximately 15–30 per cent and, allowing for losses in the system in the London area (in the open ground), the output very roughly might be about 150–300 kWh/m²y, the area used is the swept area (see Appendix B). Figure 6.11 shows a typical horizontal-axis design at a visitors' centre in a park in east London. Figure 6.12a shows some less common vertical-axis turbines and Figure 6.2b is a delightful vertical-axis Chinese windmill. Keen observers of vans and buses will know that vertical-axis Savonius rotors are used on roofs for ventilation.

In Appendix B there is a comparison drawn from a Dutch study of the advantages and disadvantages of axis position for "urban" turbines.

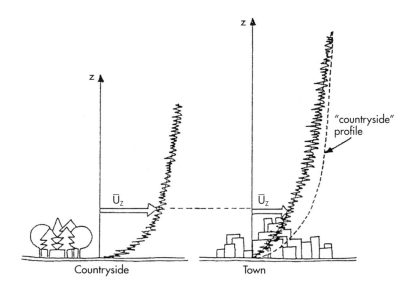

Countryside   Town

Figure 6.10
Wind speeds over varying terrains (4)

($\bar{U}_z$ means wind speed at height Z)

Figure 6.11
Horizontal-axis wind turbine at a
visitors' centre in east London

Savonius
rotor   Darrieus
rotor   H-type Darrieus
rotor   Lange
turbine

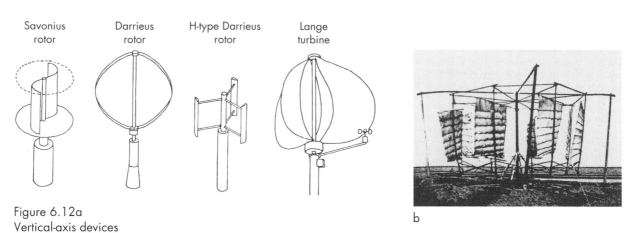

Figure 6.12a
Vertical-axis devices

b

Building type

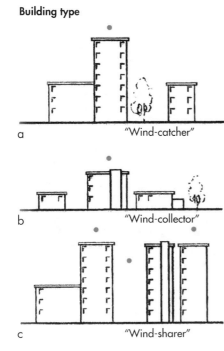

a   "Wind-catcher"

b   "Wind-collector"

c   "Wind-sharer"

Figure 6.13
Building types for urban turbines (5)

The same study examined urban building types and characterised three as
suitable for small-scale wind power; these are:

a.   the "wind-catcher" – good height plus relatively free flow;
b.   the "wind-collector" – a somewhat lower building in an area with
      more surface roughness and more turbulence;
c.   the "wind-sharer" – experiences high wind speeds and high turbulence.

Figure 6.13 compares the characteristics of these types. Obviously, the
descriptions are somewhat subjective but this research work is a valuable
step in assessing the resources available at a site.

Each urban situation will be different and so its specific potential needs to
be analysed. Figure 6.14a shows monitoring equipment to assess wind
speeds on top of the headquarters of the RIBA (at a height of 36.5 m) in
central London and Figure 6.14b shows an artist's impression of what the
wind turbines might look like on the building.

The mean wind speed is estimated to be 3.4 m/s. Initial studies indicate
that a 2.5 kW wind turbine system (with a swept area of 9.6 m²) might
provide approximately 1,200 kWh/y (6). If we compare this with our
previous estimate of electrical consumption for an energy-efficient office
building such as the BRE Environmental Building, at present about
30 kWh/m²y, we can see that 40 m² or so could be supplied by the wind
turbine. A modest contribution.

Noise levels were also assessed at the RIBA (7). The background noise
level at roof height during the day was approximately 55–65 dBA and at
night about 40–50 dBA. At 4 metres from the wind-turbine hub the noise
level was about 55 dBA at a wind speed of 5 m/s. Calculations indicated

that the RIBA wind turbine would have only a very minimal effect on urban noise levels and should pose no problems for neighbours.

What we need to develop are robust, reliable, low-maintenance and low-cost wind turbines that can easily be connected to the grid.

In the future we are likely to see combined systems of wind turbines and PV modules fully integrated with the roof design (including easy access for maintenance). Wind turbines, with or without PVs, may also be located in open spaces such as car-parks. Key questions will be those of appearance and output.

6.4 The ground

Because the ground temperature is much steadier than the air temperature (see Figure A.3) it is possible to use the ground directly as a source of cooling or indirectly as a source of heating.

Where water is trapped in an aquifer it is possible to extract it at, say, 12°C, use its cooling capacity (usually via a heat exchanger to avoid contamination) and then return it to the aquifer at a slightly higher temperature. This has been done at the BRE Environmental Building (8) and numerous other buildings throughout the country and provides an environmentally friendly way of cooling. Aquifers are common in the UK (see Appendix B) and we can expect to see this resource used more frequently in the future.

The energy available in groundwater, whether from aquifers or, say, tidal mudflats (9), or from the ground itself, can also be upgraded using heat pumps to provide a source of heat. However, most heat pumps currently available use refrigerants which deplete the ozone layer to a greater or lesser extent and/or contribute to CO_2 emissions, and so a brighter, long-term future for them awaits new materials (or the reuse of old ones, such as ammonia).

In some cities, water in aquifers is, curiously, a growing problem. This is because we are extracting less water from the aquifers for manufacturing and processing industries than we did in the past and so the water level is rising – in the case of London, at a rate of 2 m a year. Figure 6.15 is a sketch showing water levels at Trafalgar Square.

Proposals to deal with this include drilling new boreholes and reactivating existing ones. An innovative proposal includes using some of this water to cool the London Underground system (11).

Geothermal energy is also available. For example, in a number of areas in the Paris region, geothermal energy is exploited by extracting water at 73°C from an aquifer at a depth of 1500–2000 m and using it for the hot-water service (12). In the spa city of Bath the water is at a temperature of 47°C.

6.5 Community heating (CH) and combined heat and power (CHP)

In urban situations, with higher population densities and thus higher energy demands, there is the possibility of providing heat from a central boiler plant and then distributing it via a pipework system to adjacent buildings. In the past this was often known as district heating and is now more commonly known as community heating (CH). It was especially popular in modernist housing schemes in England after the Second World War. Figures 6.16a and 6.16b show housing by Lubetkin (the architect of the delightful Penguin Pool at London Zoo in Figure 6.16c) and the circular building with its tall chimney that served as boiler house and laundry (now a community meeting room).

a

b

Figure 6.14
Wind energy at the RIBA, central London
a. Monitoring equipment
b. Artist's impression

Figure 6.15
Water levels at Trafalgar Square (10)

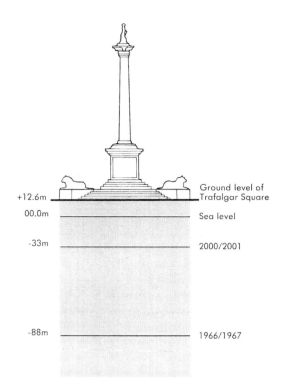

+12.6m — Ground level of Trafalgar Square

00.0m — Sea level

-33m — 2000/2001

-88m — 1966/1967

Because in cities it is likely that there will also be a high demand for electrical energy, it is common to introduce a CHP plant that produces electricity and recovers the waste heat from this process to supply part of the heating load. Often this can have a significant environmental benefit in that less CO_2 is produced than with conventional separate heat and electricity supplies. Figure 6.17 shows a group of engineers examining a CHP unit serving a housing estate for 2,600 people in Tower Hamlets, London, and providing 120kW of electricity and 250kW of heat; backup heat is provided by boilers and additional electricity comes from the grid.

CH and CHP are in successful use in a wide variety of large schemes. To cite only a few examples:

*   Sheffield burns waste to provide energy for a CH network that has primary heat sources amounting to about 30,000kW (13);
*   a number of universities, including De Montfort; Churchill College, Cambridge; University College, London; and Imperial College, London, use CHP for their facilities with excess power in some cases being exported to the grid;
*   CHP has regularly been used in a variety of large building types from hospitals to sports centres with swimming pools (14).

In addition, CH/CHP is being used at the Greenwich Millennium Village and at the Peabody BEDZED development (see Chapter 16).

Other important initiatives include Woking's development of a sustainable community-energy system in the town centre including CHP, thermal storage and absorption cooling (15). Mixed-use developments with their variety of load patterns will tend to favour CHP.

These and many other installations indicate the viability and potential of CH/CHP. Its future cost (in the year 2020) is also believed to be promising.

None the less, the case for CHP is a complex one that is continually varying with changes in technologies available; the cost of fuels, particularly

a

b

c

Figure 6. 16
Priory Green and the Penguin Pool by Lubetkin
a. View of the Priory Green estate, London
b. Boiler house and laundry at Priory Green
c. Penguin Pool, London Zoo

gas and electricity; and the CO_2 produced. Normally, a comparison is made between the efficiency of a CHP unit and the efficiency of the separate supply of heat and electricity. Figure 6.18a shows a typical comparison and it appears to be quite clear that CHP is more efficient and so the "better" choice. However, in reality in many CH/CHP schemes the CHP unit is part of a system in which backup is provided by both gas and electricity from national distribution systems. The reasons for this are technical and economic. For simplicity, if we consider a group of, say, 200 new energy-efficient homes for much of the year, no space heating at all will be required. However, there will be a year-round demand for hot-water heating. It thus makes sense to size the CHP unit on the hot-water demand so that one can maximise the number of hours the CHP unit runs (a common rule of thumb is that the unit should run at least 4,000 hours annually to be viable) rather than size it on the combined space heating and hot-water load and then have costly, unused spare capacity. The backup then comes from the national gas-distribution

Figure 6.17
CHP unit at Tower Hamlets, London

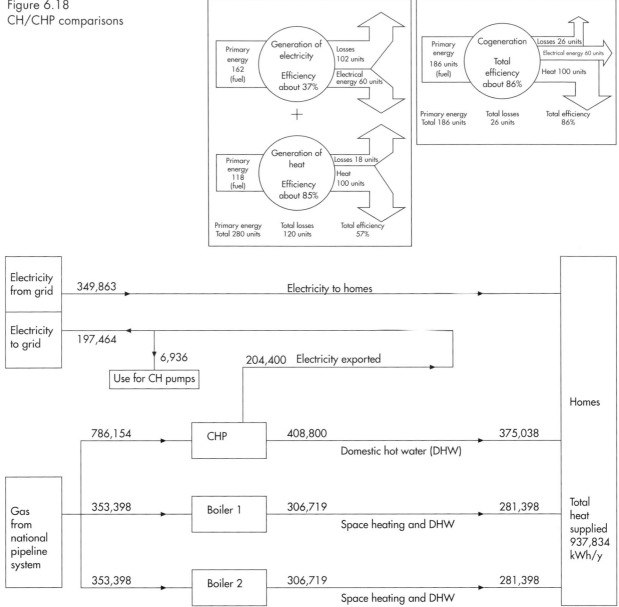

Figure 6.18
CH/CHP comparisons

system. Similarly, the unit is not based on peak electrical demand but rather uses the grid as a backup (and indeed exports electricity to the grid when it is not needed on site). Thus, for a "real" system in which perhaps 40 per cent of the heating demand and the equivalent of about 57 per cent of the electricity load is met by the CHP unit, the more appropriate comparison is between separate supplies and CH/CHP with backup from gas and electricity. The analysis of such a system is shown in Figure 6.18b; the primary energy efficiency corresponds to about 65 per cent.

A detailed analysis of the feasibility of CH/CHP was carried out for the Coopers Road housing development (Chapter 11). The four courtyards are fairly compact and the plant rooms were all located on the closest corners of the blocks to reduce the pipework linking the buildings to the main plant room with its boilers and CHP unit. The pipework connecting homes within each courtyard is extensive.

The (delivered) energy demand of the total of about 160 homes amounts to approximately 1,103,334 kWh/y for space and hot-water heating and 349,863 kWh/y for electricity (see Figure 1.6). The options considered were:

1. individual gas boilers for space and hot-water heating to each home and grid electricity;
2. CH/CHP with gas-fired boiler backup to provide space and hot-water heating. Electricity from the CHP unit to be used for the "landlord's" supply and the rest sold to the grid. Each home to have grid electricity.

The CHP (70 kW thermal, 35 kW electrical) unit was sized on the basis of supplying the hot-water load. A traditional reciprocating engine unit was selected, using availability, size and cost as the criteria.

The unit (as suggested above) will supply approximately 40 per cent of the annual total heat demand and will produce the equivalent of 57 per cent of the annual electrical demand. Compared with the option of individual boilers in each home and standard grid connections, the annual CO_2 saving is 18,000 kg. The figures assumed that individual heat meters were not included because of the high additional cost – this could be considered to be optimistic because energy use might be higher than allowed for in such a case.

The estimated additional capital cost of CH/CHP is £200,000 and so the cost per kg CO_2/y saved is £11.1. This is compared with a number of other measures in Figure 6.21. The overall economic calculations are based on electricity being sold to the local electricity supplier at a none-too-optimistic tariff. If instead the electricity (provided there was sufficient surplus) were sold directly to occupants by the system administrator, the return on capital would improve (16); such an arrangement is sometimes referred to as a private wire. None the less, the straight payback period was encouragingly estimated to be 6–7 years; this would of course be lower if a grant were received and the likelihood of such was thought to be very high.

This discussion raises quite a few issues. There is clearly an environmental advantage in lower CO_2 production but it is perhaps less than was anticipated. One reason for this is that heat losses in the pipework distribution system can be significant and the energy for pumping is also of importance. In catering there is a saying among chefs that the closer the kitchen is to the dining room, the better the food. Something similar is true here in that piping gas to individual homes and only then turning it into hot water has its merits.

What we see is that CHP can be useful but as buildings are better designed and require less and less energy, maintaining the hours of use required to make CHP efficient and cost-effective becomes more difficult. Similarly there are potential conflicts with other sources of energy. If active solar water heating is used it will compete directly with CHP for demand. Supplies chasing a demand tend to make for bad economics.

CH/CHP does also, however, have the advantage of central plant of being able to change to alternative fuels such as biomass or wastes (see below) in the future. There may be a space saving on plant but this depends very much on the design. There is perhaps less maintenance; whether maintenance should be in the hands of individuals or centralised organisations is a current subject of debate.

What is clear is that each site needs to be carefully evaluated. Also, as energy demands fall through better design, more efficient equipment and other such factors, the argument for supplying energy as close as possible to the demand seems to be strengthened. Of course, CH/CHP did this for electricity in comparison with the national grid but it was saddled, in a sense, with the problem of efficiently distributing the waste heat. There are opportunities for new technologies and this will create competition among

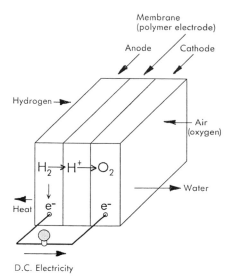

Membrane
(polymer electrode)

Anode   Cathode

Hydrogen →

Air
(oxygen)

$H_2 \rightarrow H^+ \rightarrow O_2$

Water

Heat e^- e^-

D.C. Electricity

Figure 6.19
A (simplified) PEM fuel cell

all options. For example, is it better to supply a reduced hot-water demand from CH/CHP or a very small individual boiler (yet to be developed) in conjunction with solar thermal hot-water heating? The answer is that it will depend on the analysis.

The most common CHP units are gas-fired reciprocating engines but technological advances have meant that new forms of CHP are either commercially available now or soon will be.

They are sometimes referred to as micro or mini technologies and include fuel cells, microturbines (with gas as the fuel) and Stirling engines (see below and Appendix E).

6.6   Fuel cells

There are many types of fuel cell using a variety of "fuels" but most interest is centred on cells using hydrogen, which when combined with oxygen produces electricity (d.c.) and water, giving off heat in the process, thus making these cells candidates for CHP plants. Figure 6.19 shows a schematic of a common type – the proton exchange membrane (PEM) cell. The fuel, hydrogen, enters at the anode and the oxidant at the cathode. Both anode and cathode are coated with a platinum catalyst. Between anode and cathode there is an electrolyte, which is a polymer membrane similar to Teflon. At the anode, hydrogen yields protons (H+) and electrons (e⁻). At the cathode, oxygen, protons and electrons form water.

Fuel cells were described as long ago as 1839 but the only long-term programme of research on them has been carried out by NASA, which has used them and continues to do so in space missions [17]. NASA's reasons for developing fuel cells included an assessment that they were safer than nuclear power and were able to produce potable water. Generally, their advantages are an absence of moving parts and thus very low noise levels, high efficiencies and the fact that they produce almost no pollution. Disadvantages to date have principally been cost and issues relating to the use of hydrogen as a fuel (see below).

Part of the reason for the current optimism about fuel cells is that they could be an integral part of an all-encompassing hydrogen economy (with its appropriate sources, infrastructure and users) that would link buildings and transport, solar and wind energy, urban and agricultural waste in an efficient, low-polluting, cost-efficient way. This may sound utopian but numerous initiatives are under way to form the first building-blocks of this enormous, exciting enterprise.

Figure 6.20 shows the basic elements of a scheme.

What we are likely to see is a mixture of energy sources (PVs, wind and biomass) being used concurrently.

Hydrogen, like other fuels, has its risks – it burns and is capable of forming explosive mixtures. It is not a common fuel (although during the Second World War in England buses ran on town gas, which was largely hydrogen [18], and its properties, particularly in use, will require further study to ensure that it can be introduced safely. This applies particularly to its use in transportation systems, where mobility creates more complex conditions for fuel cells than in stationary building applications. Refuelling, for example, is the subject of intensive study. Fuel cells themselves are very safe [19].

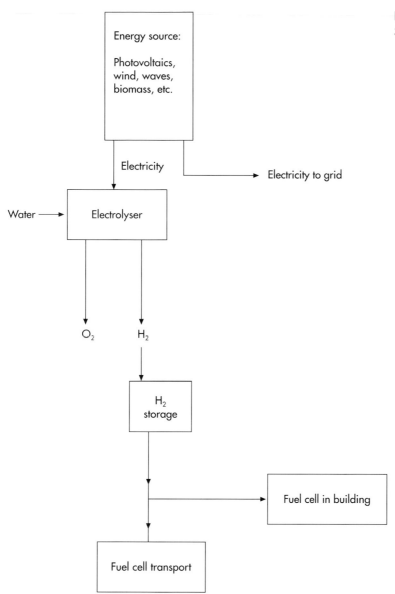

Figure 6.20
Schematic for a "hydrogen economy"

Hydrogen can be stored as a compressed gas (as in the USHER project – see section 6.11) or, for example, in special hydrogen storage alloys (20).

Regenerative fuel cells, that is, fuel cells that convert electrical energy to chemical energy and back again, are also likely to be part of the storage element of the infrastructure. The hydrogen economy with storage can help deal with the variability of the energy available from ambient energy sources such as solar and wind power. Energy from PVs, for example, could be stored to meet night-time lighting loads.

Fuel cells are beginning to make their way into our real terrestrial world. A 200 kW cell supplies electricity and power to a New York Central Park police station (21) and a similar-size unit is being installed in Woking to provide electricity and heat for a recreational centre with a pool. In Germany, a Swiss-built solid-oxide fuel cell (see Appendix E) that produces 1 kW of electricity and 3 kW of heat is being tested in individual houses, where it is installed in the basement (22). Backup heat is from a boiler and electricity can either be topped up from the grid or sent to the grid if there is a surplus. This idea of using energy (both electricity and heat) in the building so that there are no transmission losses is one of the great attractions of building-based CHP. What is less clear, though, is whether costs will be competitive with other environmentally friendly technologies. Table 6.4 shows that experts are reluctant to estimate future costs.

Fuel cells used as CHP plants for groups of buildings will tend to favour higher densities and more compact forms so that distribution losses are low. In other respects they are likely to be discrete, and if they replace traditional boilers it will be a quiet revolution. But, as always, it is a question of balance: in this case there needs to be a demand for both the electricity and the heat, and well-designed buildings and cities will require less of both.

6.7   Waste

Future sustainable waste policies are likely to include recycling (see Chapter 9), composting and recovery of energy from waste.

Waste is currently used as a source of energy in a number of large-scale incineration plants such as the Sheffield plant referred to above, and during the post-war period in England, when large housing estates were being built, engineers regularly considered incinerators as sources of heat for community heating schemes. Looking at Figure 1.6 we can see that the energy in the waste is significant. Of course, as more waste is recycled this use of the resource would reduce.

Incineration, however, is currently the subject of a lively debate, with the process being criticised particularly on grounds of design and management of the facilities, creating the possibility of the release of dioxins (23, 24, 25). How this will develop remains to be seen but alternatives are available, including pyrolysis and gasification (26).

Much waste now goes to landfill sites and in some cases the methane produced in such sites is captured and used as an energy source. Sewage sludge can also be treated in digesters to produce methane gas. This can then be burned to produce "green" electricity (27).

At the Vauban Passive House (1,600 m², 30 inhabitants) in Freiburg, Germany, human, organic kitchen and garden wastes are treated in a digester and the methane produced used as cooking gas (28).

In Malmö (see Chapter 17), a sophisticated waste strategy is used to recover energy and materials.

The entire field is one that requires constant review to keep abreast of both the economic and the environmental aspects.

6.8   Biomass

Biomass is plant material used as an energy source for either buildings or transportation. The astonishing thing about it is that we can produce it relatively quickly. If one considers that 1,000 kg of dry biomass has about the same energy content as 400 kg of crude oil (29), one can appreciate the potential. Biomass has great attractions and for some is seen as the answer to the question of how we can produce carbon-neutral buildings in the next decade or two.

Biomass is considered to be CO_2 neutral because its combustion does not result in a net increase in atmospheric CO_2; this is because such crops absorb CO_2 during photosynthesis (30). A further advantage is that when energy crops replace agricultural crops they appear to have a beneficial effect on wildlife and biodiversity (31).

Table 6.2
Energy characteristics of biomass fuels (34)

Item	Primary energy saving of biomass compared to fossil fuel kWh/ha-y	Fossil fuel replaced
Rapeseed oil	10,300	Diesel fuel
Willow	19,500	Light oil
Miscanthus	41,700	Light oil

For space heating, biomass has included beech and willow coppicing and wood chips from farms (32) or urban areas (see Chapter 16). Recently in the UK, interest has been growing in more exotic fuels such as refined vegetable oils derived from rape or sunflower seeds. Tests are under way on a grass, Miscanthus. Research is also being carried out into the use of biomass for fuel cells (33).

Table 6.2 shows the primary energy saved when biomass replaces fossil fuels.

Another detailed analysis of Miscanthus for electricity production (through gasification) gave an energy balance (energy output less energy input including transportation) of 37,252 kWh/ha (35). If we compare this with the (delivered) electrical energy demand for the Coopers Road housing, we can see that to meet the electricity demand of the 1.69 ha site would require a Miscanthus field of about 9.4 ha.

For a projected cost comparison of biomass fuels (i.e. energy crops) with other alternatives, see Table 6.4.

6.9   Ranking energy measures

So what might one do? Where should one put one's money?

Generally, one is trying to arrive at a reasonable economic balance of providing environmentally friendly energy to meet a low-energy demand. A very simple approach is to be guided by data like that in Figure 6.21.

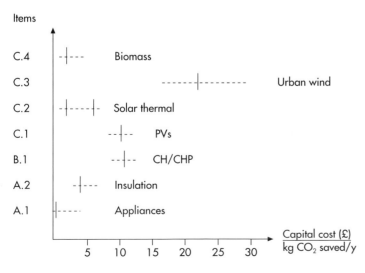

Key
A   Items that reduce energy demand
    1. Appliances
    2. Increased Insulation
B   Items that change the way energy is supplied but still use fossil fuel
    1. CH/CHP
C   Items that supply energy
    1. PVs
    2. Solar thermal
    3. Wind energy
    4. Biomass

Figure 6.21
Approximate economic costs of CO_2 reduction measures

Notes

General

Energy-saving measures will also tend to reduce radiator and boiler sizes.

20-year lifetimes are assumed throughout; the convention used is that if an item costs, say, £100,000 and it saves 10,000 kg CO_2 each year over 20 years, the capital cost per kg of CO_2 per year saved is £10. Emissions: 0.47 kg CO_2/kWh (electricity); 0.19 kg CO_2/kWh (gas).

There is a wide mixture of items that are not necessarily strictly comparable and so caution should be employed. Thus, for PVs there is a straight capital outlay but with CH/CHP it is the differential between two capital costs. The horizontal dashed lines are very rough estimates of ranges. The intersections of the vertical lines with the horizontal lines form the points referred to below.

A.1   Based on a 6.4 cubic foot refrigerator with a base energy consumption of 230 kWh/y and an improved consumption of 153 kWh/y.
Indicated point: 0.5.

A.2   Additional insulation, e.g. increase insulation from 100 mm to 150 mm and roof insulation from 150 mm to 200 mm to 300 mm.
Indicated points: 3 to 4.

B.1   Change from individual boilers and grid electricity to CH/CHP with boiler and grid backup.
Indicated point: 11.1; based on the Coopers Road site discussed in this chapter. Running costs and maintenance costs are likely to be lower with CH/CHP but as these are highly variable they have not been considered here.

C.1   Installed cost of a monocrystalline PV system: £470/m².
Assumed annual output: 100 kWh/m²y.
Assumed lifetime: 20 years.
Indicated point: 10.

C.2   The lower point of 2 is based on large solar roofs. The higher one of 6 is based on a family-size unit of approximately 4 m² of collector area at a cost of about £2,000, including hot-water cylinder and controls.

C.3   Central London, small-scale, the range is probably quite wide, say, 15 to 30 or more.
For a 2.5 kW roof-mounted wind turbine in central London:
Assumed cost: £1,300/m² of swept area.
Assumed output: 120 kWh/m² of swept area –y.
Assumed lifetime: 20 years.
Indicated point: 22.

C.4   Based on a proposal for the Cambridge University Botanic Garden (see Chapter 9). Change from a 200 kW oil boiler to 200 kW biomass boiler at an additional cost of £37,500. Maintenance and fuel costs not considered.
Indicated point: 2.

This data comes with numerous warnings: full life-cycle costs would be much more meaningful; each project needs to be evaluated on its own merits; economies of scale are possible; market costs vary constantly; and so on. None the less, it is a start; further refinements will be made with time.

Not everyone will agree with these figures – one recent analysis, for example, finds that CHP is preferable to higher insulation levels (36). Much more work is required.

These figures can be compared with large-scale estimates of the costs in the year 2010 of various CO_2 reduction strategies for the EU as shown in Table 6.3.

One is struck by how much lower these estimates are than those in Figure 6.21. This could be because the figures in Figure 6.21 are for real, current projects in (often difficult) urban settings and Table 6.3 is a projection for the year 2010. The projects may also be of a smaller scale than those envisaged in Table 6.3 and UK costs may be higher than figures applied across the EU as a whole. Very broadly, Table 6.3 argues for investment in CHP and wind power with solar thermal following and PVs the "Cinderella", of the energy ball. However, if one considers all the opportunities in the EU (or in the UK) a mix of options is the best strategy. In urban situations, the availability of roofs may make PVs a better option than wind energy. As always, each case needs to be analysed on its own merits and at the appropriate time as energy costs from these technologies vary constantly.

Only a wise man (or a group of wise men and women such as the Cabinet Office Energy Review Group) would risk predicting the cost of different energy sources in 2020. Table 6.4 shows their estimates (38)

Data like that above can be used in conjunction with an energy consumption and CO_2 production analysis to decide an appropriate environmental and economic strategy. What one needs to do is keep two curves in view, as shown in Figure 6.22.

One starts at the left of the graph and introduces energy-reduction methods. Then, as these start to become more expensive because of diminishing returns at a cross-over point, one switches to supplying energy in environmentally friendly ways and continues (until one's money runs out).

Table 6.3
Options for reducing CO_2 emissions in 2010 (37)

Option	Investment cost (£) per kg CO_2 saved/y	Ranking
Commercial CHP	0.19–0.20	1
Small scale CHP	0.27–0.27	2
Wind power	0.22–0.29	similar to 2
Solar thermal	0.78–0.88	3
PVs	1.26–1.89	4

Table 6.4
Estimated future power costs

Technology	2020 cost (pence/kWh)	Mean of range	Ranking based on mean
PV (solar)	10–16	13	7
Onshore wind	1.5–2.5	2.0	1
Offshore wind	2.0–4.0	3.0	3
Energy crops	3.0–4.0	3.5	4
Wave	3.0–6.0	4.5	6
Fuel cells	Unclear		
CHP	1.6–2.4	2.0	1
Micro CHP	2.0–3.0	2.5	2
Nuclear	3.0–4.5	3.8	5

Figure 6.22
Generalised cost curves

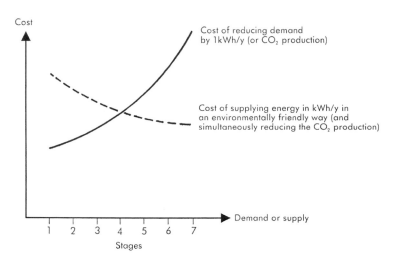

6.10   The future of energy and its design implications

In much of the twentieth century the energy supply industry was concerned with developing massive power stations that used fossil fuels or nuclear power to provide electricity via a centrally organised, highly controlled grid.

Environmental arguments now favour a transition to a solar society where renewable sources of energy provide in a more decentralised way for our needs, taking into consideration local conditions. It goes without saying that this transition should take into account economic and social concerns as a new framework is put into place. The framework needs to consider demand reduction and provision of supply simultaneously.

In this transition there are a number of unknowns. The first is whether energy will be supplied locally or remotely and the answer is a combination of both, although the proportions of each are unknown. To generate electricity remotely involves transmission losses of, say, 7 or 8 per cent. This is significant but it may be reduced in the future with technological advances such as super conducting cables. Urban space is, of course, limited and it is easier to collect energy (e.g. wind) or process materials in order to extract it (for example, waste and biomass, although this will depend somewhat on local conditions) outside the city walls.

These new sources, ranging from solar, wind, hydro (waves and tides), landfill, sewage gas, small-scale CHP, biomass and so on, are known loosely as "embedded" generators because they are local and are owned and operated by others than the national grid. In the UK there are hopes that such sources will provide 20 to 50 per cent of the electrical supply by somewhere between 2010 and 2030, depending on one's genetic allocation of optimism.

Within the city we will see a multiplicity of energy sources all competing with each other. Selection will be based on the local environment (in the best Darwinian tradition) and costs.

In some cases we are likely to see systems operating in parallel, for example fuel cells with boiler backup to allow for varying seasonal and diurnal demands for heat and power. There is likely to be a miniaturisation of components including boilers to suit the much lower loads that will occur as demands are reduced through design.

We are also likely to see transport and buildings more closely linked as in the Cambridge project described below.

And in the long term we can expect to see a shift towards a hydrogen economy with fuel cells playing a major role.

But what should poor urban designers, architects and engineers do now? Design for flexibility in this exciting but somewhat unpredictable future. Buildings are likely to see three or four different energy-supply systems in this century (as the author's nineteenth-century home has seen over a period of about 120 years). Pragmatically, this means shaping the forms and buildings to allow for solar energy, laying out the services distribution and allocating space (even if it is not constructed) on the urban site to incorporate either a centralised source or individual building sources, and allocating riser space so that services can be supplied from the ground up or the top down, for example PVs and solar thermal. Nothing could be easier.

6.11   Energy and transport

In the next decade or so we can expect to see continuing improvements in current internal combustion engine technology, resulting in improved mileage and lower levels of pollution; we'll also see more trams. However, the radical hope is that new solutions, including electrical vehicles in the first instance and then fuel-cell-powered ones, will become widely available. Such "clean", mass transport systems should help reduce problems of congestion and pollution in urban areas.

Electric cars are available now but are constrained by battery technology (see Chapter 16 for an example of electric car use).

Fuel-cell vehicles use essentially the same technology discussed above for fuel cells in buildings but with the additional constraints imposed by the need to deal with a moving fuel. Optimists hope that fuel-cell use in buildings will drive costs down through economies of scale, which then will lead to increased fuel-cell use in transport. There are the same potential advantages of being able to use renewable sources of energy and of significantly lower pollution levels.

Figure 6.23 shows a schematic of the Urban Integrated Solar-to-Hydrogen Energy Realisation Project (USHER).

USHER is proposed for Cambridge, UK, and the Swedish town of Visby. In other schemes the PV source of energy could be supplied or supplemented by electricity from, say, wind turbines or hydropower according to local conditions.

The leading fuel options are either to have an on-board reformer that takes another fuel, usually methanol, and produces hydrogen from it, or to rely on hydrogen being available directly at filling stations (40).

"Clean" vehicles are already being developed and are in use. Westminster City Council purchased a van powered by a 5 kW alkaline fuel cell, which combines hydrogen and oxygen to provide electricity (41); the oxygen is derived from air by using a chemical scrubber.

This is a field of enormous potential, which is likely to affect urban form in that PV-generated electricity could be used where required and, to a large extent, when required. Transport and building will be another link in our complex urban future. We should also be able to breathe a bit more easily because of PVs.

Figure 6.23
Solar-to-hydrogen transport (39)

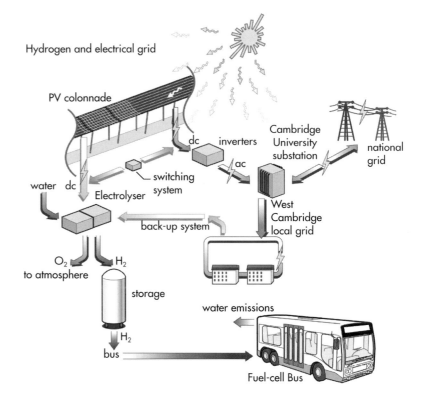

6.12 Information technology

A separate book could be written on Information Technology (IT) and the sustainable city. Very broadly, we can expect to see it blossom in at least two major areas:

1. control of systems and components;
2. provision of information to users.

Control

IT is already being employed in building engineering systems to control more closely. In a rather simplified way, this means providing energy only when it is needed, where it is needed, and in the required amount – a bit like Goldilocks and the three bears. In addition, predictive techniques for the weather will have a growing influence on urban buildings and transport systems.

IT will become more and more widespread and we are likely to see, for example, low-cost energy-collection devices that track the sun. We will also see common building materials such as concrete blocks and finishes incorporate IT (see Chapter 7), as well as components such as glazing, which will respond to changing environmental conditions.

In transport systems IT will be used to give priority to public vehicles over private uses and facilitate safer trips by bicycle or foot.

The overall result of these developments should be a more efficient use and production of energy and lower CO_2 emissions.

Provision of information to users

At the design level, detailed databases using local information will enable better-informed choices to be made about urban forms (as we saw in

Chapter 1), material selection, choice of ambient energy source and so forth. It will be possible to predict and monitor performance.

IT will enable people to participate in the design process and will inform them as users of the designs of the status of systems both artificial and natural. This will include data on air quality, energy stores, water reservoirs, the energy production of their PV panels, their electricity consumption, security at the front door and in the street, availability of seats in the next bus, the status of the shared cars, recycling results, the number of bird species in the city at any one time, and endless other items (see, for example, the very active approach being taken at Malmö (Chapter 17)).

This in turn will create a greater individual and communal awareness of our environment and how it can be improved. This role of education and stimulation will grow. IT will also allow occupants greater, and more efficient, control of their environment.

Finally, IT will change our social patterns with, for example, video-conferencing reducing the need to travel ("Why take a car if you can take a mouse?") and so will lead to environments rich in information in cities and rural areas.

6.13   Guidelines

1.   Solar design affects both urban and building form.
2.   Keep your options open. In designing the site and its buildings, allow for change. During the lifetime of the project, it will probably see at least three or four different sources of energy.
3.   Analyse demand and supply spontaneously. Reducing the energy requirements can have environmental, economic and other benefits.
4.   A mixture of renewable energy sources, from PVs to wind to biomass, is likely to be the way forward for most cities (and their surrounding regions).
5.   Community heating with combined heat and power has an important role to play in carefully selected applications.
6.   Fuel cells hold the promise of providing a pollution-free source of heat and electricity. They will be an essential part of the future hydrogen economy.
7.   Our "waste" is an important potential source of energy that should not be neglected. How it is treated has a number of impacts, including those on land use and health.
8.   Energy and transport are related and should be considered together.
9.   IT will touch every aspect of the sustainable city, from design through to the comfort and security of its citizens.

REFERENCES

1.   Butti, K. and Perlin, J. (1981), A Golden Thread. Marion Boyars, London.

2.   Mangold, D. (2001), Solar in the city. Renewable Energy World. 4 (3), pp. 100–111.

3.   Anon. (2002), Renewable Energy in the Built Environment. Building Centre Trust, London.

4.   Gandemer, J. and Guyot, A. (1983), Wind Protection – The Aerodynamics and Practical Design of Windbreaks. Project No. ED202. Building Research Station, Garston.

5. Study by Ecofys and Delft University of Technology, cited in Timmers, G. (2001), Wind Energy Comes to Town: Small Wind Turbines in the Urban Environment. Renewable Energy World, May–June, pp. 113–119.

6. Initial assessments by S. Archer, Max Fordham LLP. Private communication, February 2002.

7. Gillieron, P. (2001), RIBA – Acoustic Design Environmental Noise Survey: Wind Turbine Noise at Nearby Housing. Paul Gilleron Acoustic Design, London.

8. Thomas, R. (1999), Environmental Design. Spon Press, London.

9. Bunn, R. (1998), Ground Coupling Explained. Building Services, December, pp. 22–27.

10. Holliday, R. (2001), The man who's holding back London's tides. The Evening Standard, 2 May, p. 21.

11. Maidment, G. and Missenden, J. (2001), Sustainable Cooling Schemes for the London Underground Railway Network. CIBSE National Conference 2001 (Part 1) paper, London.

12. Habib, A. (2000), Ile-de-France: la geothermie en toute discretion. Le Monde, 15 November, p. 14.

13. Anon. (1995), Community Heating in Sheffield. BRECSU Good Practice Case Study 81. BRE, Garston.

14. Anon. (1998), Energy Efficiency in Sports and Recreation Buildings: A Guide for Owners and Energy Managers. BRECSU Guide 51. BRE, Garston.

15. Jones, A. (2001), Woking: Energy Services for the New Millennium. Woking Bureau Council, Woking.

16. Anon. (1999), Selling CHP Electricity to Tenants – Opportunities for Social Housing Landlords. DETR New Practice Report 113. BRE, Garston.

17. Fry, M. (2001), Fuelling the Future. New Scientist, 16 June 2001, No. 14, Inside Science, pp. 1–4.

18. Ibid.

19. Ibid.

20. Anon. (2001), Progress in H_2 Storage Materials. Fuel Cells Bulletin, No. 36, September, p. 7.

21. As reference 17.

22. Anon. (2001), Domestic Fuel Cells Make a German Debut. Electrical Review, 234 (19), p. 16.

23. Brown, P. (2001), Incinerator Breaches Go Unpunished. The Guardian, 22 May, p. 8.

24. Williams, A. (2001), Ashes to . . . dioxins. Architects' Journal, 2–9 August, pp. 36–37.

25. Williams, A. (2002), Brimful of ash. Architects' Journal, 10 January, p. 48.

26. Ling, M. (2001), Turning household waste into light and power. EIBI, May, p. 14.

27. Holden, R. (2001), Mogden STW – Sludge Stream Improvements. Water and Sewerage Journal, Issue 3, pp. 35–37.

28. See the website www.vauban.de

29. Bullard, M. (2001), Economics of Miscanthus. In Jones, M. and Walsh, M., Miscanthus for Energy and Fibre. James and James, London, p. 164.

30. Santos Oliveira, J. F. (2001), Environmental Aspects of Miscanthus Production. In Jones, M. and Walsh, M., Miscanthus for Energy and Fibre. James and James, London, pp. 172–178.

31. Ibid., p. 176.

32. Ashley, S. (1999), Branch lines to Weabley. Building Services, February, pp. 32 34.

33. Lehmann, A., Rusell, A. and Hoogers, G. (2001), Fuel cell power from biogas. Renewable Energy World, November–December, pp. 77–85.

34. Kaltschmitt, M., Reinhardt, G. A. and Stelzer, T. (1996), LCA of Biofuels Under Different Environmental Aspects. In Chartier et al., Biomass for Energy and the Environment – Proceedings of the 9th European Bioenergy Conference, Copenhagen. Elsevier, Oxford.

35. As reference 30, p. 172.

36. Orchard, W. (2001), The case for CHP. Building Services, October, pp. 45–48.

37. Wilberforce, R. (2000), Furthering understanding. International Glass Review, Issue 3, pp. 28–30.

38. Anon. (2002), The Energy Review; A Performance and Innovation Unit Report. Cabinet Office Energy Review, London.

39. Lloyd, T. (2001), Usher shows the way. Electrical Review, 234 (21), p. 28.

40. Wilks, N. (2001), Vapour trials. Professional Engineering, 13 June, pp. 30–31.

41. Anon. (2000), Westminster buys its first fuel cell vehicle. Energy World, April.

FURTHER READING

Anon. (undated), Solar Solutions. Solar Century, London.

BRITS, the newsletter of the British Photovoltaic Association. Website: www.pv.uk.org.uk

Daniels, K. (1998), Low-Tech Light-Tech High-Tech. Birkhauser, Basel.

Evans, R. J. D. and Frost, J. C. (1998), Fuel cells coming of age for both transport and stationary power. Energy World, No. 255, pp. 8–11.

Karvountzi, G. C., Themelis, N. J. and Modi, V. (2001), Maximum Distance to which Cogenerated Heat can be Economically Distributed in an Urban Community. Poster paper 4528 at the 2002 ASHRAE Winter Meeting.

Lloyd-Jones, D. (2000), Doxford International, ETSU Report S/P2/00275/REP. ETSU for the DTI.

WEBSITES

Construction Resources: www.ecoconstruct.com

European PV city guide: http://pvcityguide.energyprojects.net/

Solar Century: www.solarcentury.co.uk

Solartwin Solar Water Heaters: www.solartwin.com

7

Materials

Randall Thomas

7.1 Introduction

Le Corbusier wrote that "the old city dies, and the new city rises on its ruins" after climbing the Empire State Building in 1935 (1). Whether or not one agrees with his somewhat contentious statement, what is true is that the unending creation of our urban environments entails huge material flows to and from the city. In the UK, approximately 6 tonnes of aggregate is required per person every year for roads, buildings, farms and for de-acidifying rivers (2)! Currently, 2 tonnes of concrete (half of it with reinforcing steel) is used per person annually (3).

The good news is that there are a number of ways for us to construct more environmentally friendly cities than we have done in the past; the bad news is that we can't yet agree which materials are ideal for the task. The area is a minefield of claim and counter-claim. What we have tried to do below is set out the main lines of the options. As always in this field each situation is different and one should perhaps concentrate on what the appropriate materials are for a development rather than on what the illusory ideal ones might be.

Our cities need materials for buildings, roads, services infrastructure (drains, pipes, cables and so forth) and the landscape. What is striking is that a rather small number of materials account for most of the energy consumption and hence the CO_2 production of our built environment. For a typical house, for example, approximately 75 per cent of the total energy consumed in producing the building materials is for concrete, plasterboard, bricks and mortar; glass, steel, copper and paint account for 13 per cent and timber for 8 per cent (4, 5).

7.2 Selection of materials

What criteria might one use to choose materials? There is some agreement that the following are of assistance (6):

1. impact on the global environment, e.g. CO_2 emissions, destruction of the ozone layer;
2. impact on the local environment of sourcing the material, e.g. felling, quarrying (Figure 7.1 shows the effect on a low mountain range just 40 km north of New York City of quarrying for rock for the city's skyscrapers and roads);
3. impact on the local or global environment of processing the material;
4. embodied energy content – the energy used in sourcing, transporting and processing (the Millennium Village in Greenwich, for example, originally set out to achieve a 50 per cent reduction in embodied energy in its materials);
5. health hazards associated with processing, fabricating or preserving materials (similarly, materials should be examined for their effect on the health of the building's occupants);

Figure 7.1
Quarry in Rockland County, New York

6. life expectancy of the material and its potential for reuse or future recycling.

What one needs to do is to view how the materials will be used and what the effect will be over the entire lifetime of the development – easier said than done. Another consideration is that there are different ways of looking at the criteria; for example, a long-life material may in fact discourage its substitution with a more advanced technology. One sees this possibility with photovoltaic roofs where one might want current roofing materials to last no longer than the year in which PVs become cost effective.

It is relatively easy to set out the embodied energy content of materials as in Table G.1 (Appendix G) but this is only part of the issue. If two materials are being compared, one needs to examine how much of each is required to perform the same function. Or, indeed, since a material may perform two or more functions, for example concrete as a structural material and as a means of providing thermal mass, both functions.

Currently, the embodied energy in a typical office building is about 7–10 per cent of the total energy used over the lifetime of the building (7, 8). However, as running costs are reduced, the significance of the embodied energy increases.

A number of specialised works deal with selection of materials (9, 10). Here we will simply make a few brief comments.

Biological materials

Two of the three pigs, indeed, used biological materials for their homes and since the time of their initial experiments with the wolf, durability has increased because the technologies have improved.

Wood

Wood is often seen as one of the most environmentally friendly materials. It is clearly a renewable resource and some of it can be sourced locally. However, there tends to be a mismatch between where timber is grown and where it is consumed in the world as a whole and so issues of environmental impact and embodied energy can be exceptionally important. The UK, for example, imports 90 per cent of its timber demand (11).

Areas of development include the use of less-known timber technologies, such as timber I-beams, and reducing waste in construction and demolition work. The use of insulation materials of low-embodied energy made from recycled newspaper is growing.

Preservation methods for timber need careful consideration as toxicity can be an issue. Certification systems are an area of concern for a number of environmentally concerned specifiers who have reservations about the accuracy of the information being provided.

The maximum height of timber-frame buildings is obviously limited. Five-storey buildings are common but for anything much taller we are likely to be left with a choice between steel or masonry construction.

Other materials

Agricultural wastes have found their way into buildings for years. Straw boards and now straw itself is arousing some interest. Sheep's wool (see Figure 7.2) is being introduced as an insulation material and is claimed to have a low embodied-energy content.

Concrete

It is hard to imagine many of today's cities (both buildings and infrastructure) and much of modern architecture without concrete. Interestingly, concrete may have been important in the past also – a recent theory being examined by French scientists is that the Egyptian Pyramids were in fact built from concrete blocks moulded on the spot, stone by stone and layer by layer (12).

Concrete has proved to be a versatile material and has advantages of providing thermal mass (see Figure 7.3, of the BRE Environmental Building where the floor/ceiling construction combined thermal mass and the ventilation path (13)) and acoustic insulation. The key ingredients are cement, water and aggregates (sand and gravel or crushed rock). The impact on the environment of acquiring the aggregates and the embodied energy of the cement, the most energy-intensive ingredient, have posed some concern. Cement for concrete accounts for about 2 per cent of the UK's carbon dioxide emissions (14). Embodied energy can be expressed as kWh/m^2 of floor area for a finished construction (as an alternative to kWh/m^3 of material) and, according to the British Cement Association, reinforced-concrete construction requires $417–695\,kWh/m^2$ for a range of typical structures whereas structural steel alternatives require $723–806\,kWh/m^2$ (15). The Steel Construction Institute, on the other hand, reports that recent research carried out by them for offices indicates that "there are no significant embodied energy/carbon dioxide differences between the alternative construction types" (16).

So let the buyer beware and ensure that he or she does all research thoroughly and tries to set out clearly the reasons for deciding what the appropriate material might be. Such an approach was used on an element-by-element basis for the café at the Earth Centre, Doncaster, where, for example, concrete, steel and timber were considered to be the options for the frame (17). The decision made was to use precast concrete and the justifications given were that the fire regulations required concrete floors and that concrete's thermal capacity was better than that of steel. The designers further concluded that recycled aggregates should be used.

The concrete industry is aware of the need to continue to reduce the energy inputs. Work on lowering the cement content by replacing it with recycled materials is ongoing. Pulverised fuel ash (PFA) from coal-burning power stations has been used for some time and researchers are looking at materials such as ground brick dust (18).

Metals

Although the embodied-energy content of metals is high, this can be acceptable when metals are clearly the material that best suits the function. This could be the case, for example, for the frame of a tall building. It is certainly true at present (and for the conceivable future) for cables that conduct electricity.

The embodied-energy content of materials falls with recycling. Commercial aluminium (30 per cent recycled content), for example, has an energy content that is about one-quarter less than that of primary aluminium.

Figure 7.2
Sheep's-wool insulation and a sheep

Figure 7.3
Concrete construction at the Environmental Building, showing guiding arrows for the incoming air

Often, unfortunately, choice of materials is not simple. For example, a detailed French study comparing zinc and uPVC for something as mundane as rainwater pipes showed that zinc had a lower embodied-energy content than uPVC and was chlorine free but involved cadmium in its production (19).

Glass

Glass has marked modern architecture from the Crystal Palace to Philip Johnson's Glass Pavillion to Jean Nouvel's Cartier Foundation (see Figure 5.12). Will it characterise this century's sustainable architecture? Yes, certainly in part. The energy is not that great (very roughly, the embodied energy in $1\,m^2$ of south-facing glazing is recovered by the passive solar gain in October in about a week), U-values will improve, "intelligent" materials with memories and control functions will be developed, and special glazing incorporating energy production such as PVs will become more widespread. We can expect to see glass used for its beauty and, for example, to reduce the need for artificial lighting. Glass in combination with movable insulation panels for, say, night use also has a promising future.

Material avoidance

While one side of the coin is material selection, the other is material avoidance. A modest proposal is to avoid materials that damage the environment. These include any with ozone-depleting chemicals such as CFCs and HCFCs. In landscaping, the use of peat is being avoided because of its environmental damage at the sources of extraction. Tropical hardwoods from non-sustainable sources (which covers most of them) should be avoided.

There is an important debate currently taking place about chlorine, or, more precisely, the group of substances known as organochlorines, in the environment. For many, these substances "compromise the environment as a safe place for maintaining the fertility and abundance of life" (20).

In the built environment, uPVC windows and PVC insulation for electrical cables are two examples of the use of chlorine. Alternatives are available. To quote again the author of the above comment, "Readers must judge for themselves how we can best attain a sustainable future" (21). Chlorine should be kept under review more carefully than ever.

7.3   Lifecycle considerations

It is useful to consider the lifecycle of materials in designing and constructing the urban environment. To start with, perhaps the design can use fewer materials, for example through a better understanding of the properties of the materials or through composite materials. ("Use thought rather than materials" is the catch-phrase.) There may also be an advantage in energy and CO_2 terms if materials are resourced locally (see Chapter 16). Reducing waste on site is a key issue for the construction industry. One way to do this is to construct building components in a more controlled way in off-site factories.

The recycling of materials is important. From the outset of the design, one should consider how the buildings will be "deconstructed". In London, for example, there is a growing problem in some areas of the city in finding space for new concrete piles when buildings are demolished and new ones erected (22). Existing piles cannot economically be removed so new piles need to be put in the (diminishing) space between the old ones.

With concrete construction one will want to recycle, recovering the steel reinforcement and reusing the concrete as aggregate. The BRE Environmental Building was constructed on the site of a building that was demolished – 96 per cent (by volume) of its materials (bricks, concrete, timber, copper cables, etc.) were recycled (23).

It is likely that we will move towards more and more recycling at an urban or perhaps a regional level (because of issues of dust and noise) and that the energy for these processes will come from environmentally friendly sources.

7.4   The future

Some aspects of the future can be considered now. Our buildings should incorporate a strategy for upgrading that is drawn out by the architect. This could include provision for higher insulation levels, the replacement of windows with higher-performance units in 10–20 years' time, and so forth.

Our palette of materials is widening and a number of approaches are already or will soon be available. They include:

- aerogels (sometimes referred to as "solid smoke") – ultra-light materials that are about 99 per cent air and are highly effective insulators;
- paints that change their thermal properties, for example becoming more reflective, at higher temperatures;
- translucent concrete that will allow us to think differently about light and structure;
- glazing that reflects light towards the ceiling and back into a room to provide better use of daylighting;
- greater recycling – crushed glass may be used as concrete aggregate;
- New, more efficient photovoltaic materials (both silicon and non-silicon based);
- more efficient lighting including broad-spectrum light-emitting diodes;
- new and more efficient storage materials for energy and for hydrogen.

Other materials are further away but still likely. These include:

- materials, from insulation to concrete to glazing, that combine intelligence elements (sensors, controls and communication devices) to report and change status, for example to reduce energy consumption;
- biologically based plastics – so that in the future we may grow everything, from the ubiquitous plastic bag to translucent sheets for buildings.

7.5   Guidelines

1. Materials should be selected with the environment and health among the criteria; many materials can be used to construct a sustainable future.
2. The lifecycle of materials is a key issue. Use recycled materials where appropriate and design to facilitate recycling.
3. Design should allow for improvement in materials and component replacement.

REFERENCES

1. Cited in Bilger, B. (2000), The Crumbling Skyline. The New Yorker, 4 December pp. 65–71.

2. McAlpine, E., quoted in "How blighted is my valley", The Observer, 1 April 1990, p. 73.

3. Anon. (2000), Ecoconcrete. British Cement Council, Crowthorne, Berkshire.

4. Ansell, M. (1995), Good Wood. New Builder, 7/14 July, p. 13.

5. Thomas, R. (2000), Environmental Design, E&FN Spon, London, p. 70.

6. Turrent, D. (1995), Green Building Materials, June, p. 37.

7. Howard, N. and Sutcliffe, H. (1993), Embodied energy: the significance of fitting out offices. BRE 180/22/9, Garston.

8. See reference 3, p. 3.

9. Anderson, J. and Howard, N. (2000), The Green Guide to Housing Specification. BRE, Garston.

10. Wooley, T., Kimmins, S., Harrison, P. and Harrison, R. (1997), Green Building Handbook. E&FN Spon, London.

11. Evans, B. (1995), Timber – not green enough? Architects' Journal, 4 May, p. 43.

12. Webster, P. (2001), Why De Mille didn't need all those slaves on screen. The Observer, 30 December, p. 20.

13. See reference 5, p. 202.

14. See reference 3, p. 13.

15. Ibid.

16. Anon. (1998), Talking Steel. Building Design, 23 January 1998, p. 33.

17. Demetri, G. (1997), World service. Building, 21 November, pp. 75–79.

18. Littlewood, J. (2002), Recycling concrete – a hard grind. Building, 267 (01), p. 33.

19. Anon. (2001), Bâtiments Publics et Haute Qualité Environnementale. Manuel du Stagiaire, ADEME, Paris.

20. Collins, T. (2000), A call for a chlorine sunset. Nature, 406, 6 July, pp. 17–18.

21. Ibid, p. 18.

22. Tant, R. (2001), Sustainable Development: the parts we cannot reach. RSA Journal, 2/4, p. 2.

23. Hobbs, G. and Collins, R. (1997), Demonstration of reuse and recycling of materials: BRE energy efficient office of the future. IP3/97, BRE, Garston.

8

Water

Randall Thomas

8.1 Introduction

Urban water systems are closely related to their surrounding regions. In medieval Cambridge, plentiful supplies of fresh water were brought in from several kilometres outside of the city via lead pipes under the directions of the Grey Friars (1).

London's present-day water supply comes from groundwater (see below), rivers (including the Thames) and from storage reservoirs in the surrounding area (2). London (and the Thames Valley region), with a high population but relatively low rainfall (varying from 800 mm/y in the west to 500 mm/y at the Essex coast), is representative of urban areas that need to manage precious water resources carefully.

A typical household in the UK will use approximately 128 m³/y of water (based on 2.34 persons per household) (3).

The starting point in examining the issue of sustainability is the usual one — can demand be reduced? Another important point is the relationship of the quality of the product to its intended use — for example, drinking-water quality is not needed for flushing WCs. We will return to both but first, a few definitions. Mains water is that supplied from the distribution system and, as the principal supply of drinking (or potable) water, is subject to strict quality controls — this normally involves treatment, for example, filtration, ozonation and chlorination. Approximately, 30 per cent of the mains water supply in England and Wales comes from groundwater and 70 per cent from surface water (4). Groundwater is normally abstracted from boreholes into aquifers (permeable formations from which water can be pumped). (A map of the principal aquifers in the south-east of the UK is shown in Appendix B; for the importance of groundwater for cooling in buildings, see Chapter 6.)

Treatment of both groundwater and surface water is important for sustainable cities. Groundwater, for example, can be contaminated by many processes, varying from de-icing chemicals used on roads, to acid rain resulting from combustion processes, to chloride and ammonia resulting from landfill (5). In many urban areas there is also a history of land contamination leaching into the groundwater, so this needs consideration.

As discussed in Chapter 4, how to deal with surface water is rapidly becoming a field of intense activity and importance. The tendency of urbanisation has been to seal more and more of the surface of the earth, usually with an unattractive material known as Tarmac. One result of this has been the need to deal with the run-off of large volumes of water. Another consequence of so much Tarmac has been an increase in the amount of solar radiation absorbed, thus contributing to the urban heat island. An alternative or, better, complementary approach is landscaping.

8.2 Water and the landscape

A black, impermeable surface will absorb none of the rain that falls on it and very roughly 90 per cent of the incident solar radiation. A park with trees and grass will take in almost all of the rain and most of the solar radiation. The exact amount of the sun's energy absorbed will vary with the ratio of trees to grass and the species involved but, very roughly, the figure might be 80 per cent (6).

A judicious selection of plants should mean that rainfall alone will suffice for the park provided that one can accept the grass turning a straw-yellow in dry summers, as it does regularly in particularly dry areas of the UK.

To consider a typical summer (from around May to August), a London park will have an average rainfall of 1.6 mm/day. Much of this will be used for evapotranspiration by the plants, resulting in the air being cooled and a feeling of greater comfort in the city.

Water in lakes, ponds and fountains can have the same physical cooling effect, in addition there is often an accompanying psychological effect and an element of sheer delight. Horizontal water surfaces tend to calm and laughing fountains excite – the city has a need for both. Water also offers incomparable opportunities for plant and wildlife habitats and so greater biodiversity.

At the Alhambra, water from the Sierra Nevada mountains is used to great effect (see Figure 8.1) in these and many other ways. Evaporation from the pool on a hot summer's day in Granada lowers the temperature of both the air and the surrounding buildings.

Figure 8.1
Courtyard pool at the Alhambra, Granada

8.3   Water and buildings

The mains water supply is of course used (and misused) in thousands of ways. However, the choices are in fact up to us and we can reduce consumption if we wish to use our resources wisely. In Japan local authorities encourage recycling. For example, in Tokyo, for a building with a floor area of over 30,000 m² to get planning permission it must recycle rainwater and have an in-building grey-water treatment system (7). Definitions vary somewhat but often rainwater is considered as one category, grey water as another and black water as a third. In this definition, grey water is considered to be all waste water from domestic appliances with the exception of toilets. Thus it includes discharges from kitchen sinks, washroom basins, baths, showers, washing machines and dishwashers. Black water is all water that combines into the foul drain and then into the sewer system. It thus includes all grey water and the waste water from toilets (8). (An alternative definition is given in Reference 9.)

If we look at household use, Figure 8.2 shows some typical present-day figures. Note that these are national averages and so can be significantly improved on.

Note that only 6 per cent or so is used for drinking and cooking even though all domestic water is to international standards for potable water. It is a bit like using our highest source of energy, electricity, for all power and heating in the home when one has a lower-grade source, gas, available. Note also that about one-third is used for WC flushing and that clearly this water could be of a lower quality. Thus, one must ask the question of whether there is a better way of meeting our water needs.

But, as mentioned before, the starting point is to reduce demand. This can be done in a number of ways, including the use of more efficient dishwashers and washing machines, low-water-volume WCs, taps that aerate the water stream and low-flow showerheads. By actively managing water use it should be possible to reduce consumption to a figure closer to 100 litres/person/day (this is the basis for the water use in Figure 1.6). More radical alternatives are also available, such as composting toilets (see Chapter 9) and waterless urinals.

Figure 8.2
Typical domestic water use (10)

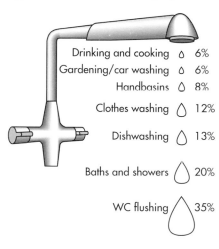

Drinking and cooking   6%
Gardening/car washing   6%
Handbasins   8%
Clothes washing   12%
Dishwashing   13%
Baths and showers   20%
WC flushing   35%

The next step is to examine how rainwater and grey water might be used to save water.

8.4   Water harvesting and recycling

The first candidate for use is rainwater. Rainwater is generally considered to be cleaner than grey water and to entail less risk of infection in the event of systems not operating properly (11). Its quality is generally good but it is contaminated to some extent by the gases, dust and living organisms in the atmosphere and by pollutants on the collection surfaces (12). Nitrogen and oxygen are the main gaseous impurities in rainwater; in and near urban areas the emissions of flue gases with CO_2 and sulphur dioxide can result in water of increased acidity and corrosiveness (13).

Traditionally, rainwater has been stored in a butt and used for watering the garden. The next likely step is to collect the rainwater, filter it and store it in, say, a basement or underground storage tank of concrete or plastic. Such systems are commercially available. Figure 8.3 shows one that has been widely installed in German homes, where the filtered rainwater is used for WC flushing, the washing machine and for the garden. Although not shown in the diagram, an automatic mains backup system ensures there is a supply in the event of insufficient rainfall.

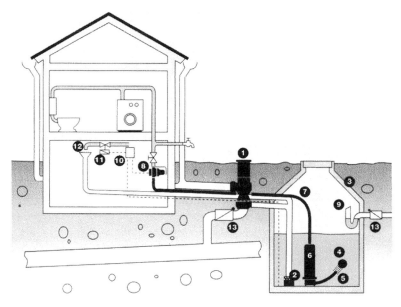

Figure 8.3
Rainwater collection and recycling system (14)

1 Vortex underground filter
2 Inflow smoothing filter
3 Tank
4 Floating fine suction filter
5 Suction hose
6 Multigo pressure pump
7 Pressure hose
8 Automatic pump control
9 Trapped overflow
10 Automatic mains back up
11 Solenoid valve
12 Type 'A' air gap
13 Anti-surcharge

The size of the tank will vary with the demand, the available area of collection surface and the percentage of demand that is to be met from the tank. (Obviously, a huge tank will protect against severe droughts but will have a cost penalty and other disadvantages.)

The cost of such a system for a home is approximately £1,850 and at current water prices (including sewerage charges) would have an economic payback period of about 50 years. The environmental benefits, however, are substantial in that the need for mains water is reduced and the infrastructure demands on both supply and disposal are reduced.

Larger-scale systems are also in existence. A group of eight single-family dwellings completed in Marburg, Germany recycled rainwater from part of their roofs and from a nearby apartment block and stored it in a $24\,m^3$ underground storage tank from which it was pumped for WC flushing (15).

Grey-water recycling is the next step – but which grey water? The use of grey water is still in the development stage and research on the performance of such systems is ongoing. Bath water and shower water are often considered the next most likely candidates for recycling; kitchen water is less valued because it contains fats and organic materials. Commercial systems that collect bath water, disinfect it and then recycle it for toilet flushing are available and research projects are under way to see if grey water can be recycled without the use of disinfectants (16).

Larger projects often provide scope for significant savings in water use. In a scheme design for a Visitors' Centre for the Cambridge University Botanic Garden, in keeping with the client's (and the consultants') very high environmental standards, water and waste management were considered from the outset. This subject is discussed in the next chapter.

Even larger projects have even more sophisticated systems. Figure 8.4 shows the schematic of recycled water used for WC and urinal flushing at the Millennium Dome in Greenwich, London.

Figure 8.4
Water recycling at the Millennium Dome (17)

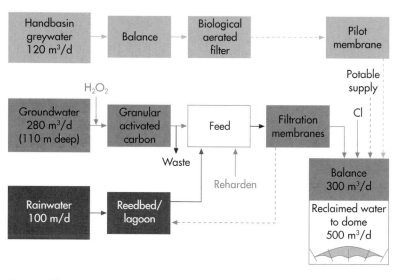

Three different sources – rainwater, groundwater and grey water from handbasins – each with a separate treatment system, were used to provide 500,000 litres/day.

Sustainable urban drainage (see Chapters 4 and 9) is essential and complements the measures discussed above.

8.5   Heat recovery on waste water

In principle, heat recovery on warm waste water is possible and the subject has been examined (18) and occasionally incorporated in projects. Maintenance is required in some designs to keep the heat exchangers free from fouling but new designs with simpler heat exchangers are being developed.

8.6   Guidelines

1.   Reduce demand.
2.   Try to ensure that the quality of the water is as high as required for the use but no higher.
3.   Organise the site so that as much of the rainwater falling onto it as possible can be used sensibly.
4.   Even if water recycling is not incorporated at the outset of a project, allow for its incorporation (in terms of space for storage, pipes, etc.) at a later date.
5.   Consider rainwater recycling and bath-water recycling for each new project.
6.   Choose vegetation that doesn't need irrigation in the summer.
7.   Water is physically and psychically nourishing – design with it.

REFERENCES

1. Trubshaw (1988), Bringing Fresh Water Across Town. Cambridge Weekly News, 24 November, p. 17.

2. Sumbler, M. G. (1996), British Regional Geology: London and the Thames Valley. HMSO, London, p. 147.

3. Crowhurst, D., Bennets, R. and Runacres, P. (2001), The Movement for Innovation: Sustainability Working Group Report. BRE, Watford.

4. Cooper, E. (1990), Utilisation of Groundwater in England and Wales. Water and Sewerage, pp. 51–54.

5. Ibid., p. 52.

6. Monteith, J. L. (1973), Principles of Environmental Physics. Edward Arnold, London, pp. 65–67.

7. Stephenson, T. (2001), Mission: Tokyo. Water Services, April, pp. 32–33.

8. Garrett, P. (1998), Water Utility Week, 14 August 1998, pp. 12–13.

9. The DETR defines grey water as water from buildings that can be reused and goes on to say "light grey water is rain water collected from roofs and used for toilet flushing and non-drinking water applications. Darker grey water is from sinks and baths which can be used for watering plants but would require extensive processing for other uses." Anon. (1998). Sustainable Development: Opportunities for Change Sustainable Construction. DETR, London.

10. Rawlings, R. (1999), Environmental Rules of Thumb Technical Note TN12/99. BSRIA, Bracknell.

11 Anon. (2001), Liquid Asset: Recent Developments In Water Savings. Eco Tech, 4, pp 34–38.

12. Treanor, A. (1980), Water Treatment, Specification. The Architectural Press, London, pp. 2–158, pp. 2–163.

13. Ibid.

14. Rainwater systems available from Construction Resources, 16 Guildford Street, London SE1 OHS. 18 (4), pp. 27–28.

15. van Zoelen, M. and Boonstra, C. (1996), A Silver Lining. Building Services, April, pp. 27–28.

16. As reference 8.

17. Palmer, J. (1999), Sustainable for a Year? Building Services, 21 (4), pp. 50–51.

18. Smith, I. E. (1977), The Recovery and Utilisation of Heat from Waste Water in the Home. School of Mechanical Engineering, Cranfield Institute of Technology, Bedford.

FURTHER READING

Grey, R. and Mustow, S. (1997), Greywater and Rainwater Systems: Recommended UK Requirements, Final Report 13034/2. Building Services Research and Information Association, Bracknell, UK.

Waste and resource

Adam Ritchie

Figure 9.1
Uncollected rubbish in Leicester Square during the 1978–1979 "winter of discontent"

9.1   Introduction

Regular refuse collection and municipal drainage infrastructure can engender a certain complacency about the quantity of waste we produce and its method of treatment and disposal. A blocked drain or even a short cessation of refuse collections serves to illustrate our dependence on a waste disposal infrastructure. Figure 9.1 shows a London park piled high with refuse during labour strikes in the 1978–1979 "winter of discontent".

Our waste is a rich mix of materials and minerals; indeed the waste industry is likely to find itself at the heart of the carbon debate because of the energy content in waste. Presently 54 per cent of industrial and commercial waste and 83 per cent of household waste goes to landfill (1) at a cost approaching EU levels of £100 per tonne (2). Clearly we must develop new waste strategies, including recycling, composting and energy recovery, and put our waste to use as a resource.

9.2   Waste in the city and in the regions – an overview

Waste in the context of the urban environment is in fact a wider regional issue. Waste generated in cities is, for the most part, transported to the surrounding regions where cheaper land prices favour landfill or large incineration facilities. Sewage-treatment works employ sophisticated and energy-intensive techniques to treat large volumes of waste water using relatively small amounts of land. These tend to be located in areas where potential odour and visual impact are lesser concerns, more typically found in the regions.

A broad picture of the current situation in England and Wales is outlined in the Waste Strategy 2000 (3), a UK Government plan for improving the management of waste and resource in the future.

Approximately 400 million tonnes of waste are produced per year in England and Wales and this amount is increasing, on average at a rate of 3 per cent per year (4). One-quarter of UK waste production can be assigned to industry, commerce and households. The remainder is constituted of construction and demolition waste, sewerage sludge and mining and agricultural waste.

Each UK household produced an average 23.2 kg of domestic waste, excluding waste water, per week in 1999/2000 (5). Figure 9.2 shows the variation in household waste output between metropolitan and non-metropolitan districts and in comparison with London. At an average of 2.3 persons per household (6), we each produce about 525 kg/person/year, which is in excess of average European waste outputs of 200–500 kg/person/year (7).

There is a philosophical question as to the level (local, city, region, etc.) at which waste should be dealt with. This is a live issue and one that is likely

Figure 9.2
Household waste by outlet and authority type

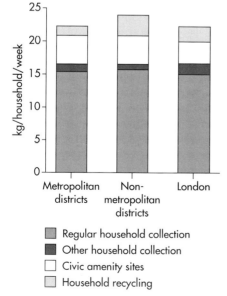

to lead to different solutions in different communities. As with energy (Chapter 6), some solutions will involve a combination of several approaches.

The key element to sustainable urban waste management is to reduce the amount of waste generated in the first place or, to quote Buckminster Fuller, to be "doing more with less". Chapter 8 examines ways to conserve and recycle water in buildings by using low-water-use appliances or grey-water recycling. A comprehensive review of products available on the market can be found in Reference 8. By lowering our water use at the outset, the quantity of waste water requiring transportation and treatment is also reduced. This illustrates the degree to which sustainable design should be approached holistically rather than by examining isolated systems.

9.3   Waste as a resource

At one level, resources flow through the environment and through the cities and buildings in which we live and work. To date there has been a linear process of exploitation, conversion, use and disposal. A sustainable philosophy involves, to some extent, turning this linear process into a cyclic one. What is waste to one person can be a valuable resource to another. Barriers such as location have tended to discourage the uptake of these mutually beneficial relationships.

Energy from waste

It is possible to define a hierarchy of waste. A proportion of household waste can be recycled in the form of materials, some wastes naturally degrade and some wastes will burn readily without additional fuel. The energy available for release from a material when combusted is described as its calorific value. The calorific values of some waste products are shown in Figure 9.3.

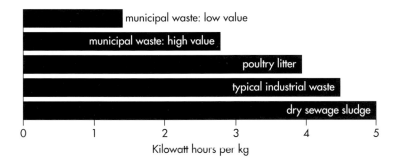

Figure 9.3
Calorific values of waste

There is a multitude of processes (direct combustion, pyrolysis, digestion, fermentation, etc.) for the conversion of waste streams into heat and/or electricity or a carbon fuel. Examples already exist at regional, city and local levels and some were discussed in Chapter 6.

At different building scales, each has its own merits and, generally, opinions vary. Individual sites will require careful consideration of suitability and context.

Recycling

Recycling of waste can be environmentally preferable to incineration, for example, if the energy saved exceeds the recoverable energy (typically 20 per cent for incineration plant (9)) plus the energy used in separating the

Figure 9.4
Multi-compartment under-cupboard bin

waste. Concerns regarding dioxin emissions from incineration plant along with the tangible benefits of recycling, such as job creation, must also be included in the equation.

Building design can facilitate the sorting process so that each form of waste enters an appropriate waste stream and is put to its most beneficial use. Allowing space for additional or multi-compartment sorting bins (see Figure 9.4) will enable households to separate easily waste at source. The Malmö BoO1 project, described further in Chapter 17, has taken this idea several steps further by providing a vacuum-tube waste transport system. This allows householders to send organic and mixed waste to central facilities for treatment and avoids the need to bring a refuse truck into the neighbourhood.

Similarly, designing separate drainage systems for surface, grey and foul (black) water is an approach to sorting waste at source, although various issues need to be addressed, some of which were discussed in Chapter 8.

Human waste

Human waste contains organisms that can cause diseases. Any form of human waste treatment must deal with these pathogens and reduce them to safe levels (opinions vary on the definition of safe (10)). Treatment must also provide the micro-organisms that break down organic matter with sufficient oxygen without robbing it from other ecosystems.

A variety of solutions exists at local and regional levels and some of these are explored below in an urban context. The present urban solution, large-scale sewage treatment works, relies on a complex sewerage infrastructure consisting of pipes, access chambers and pumps, all requiring regular maintenance. It is because of the distances human waste must travel that the predominant use of water in today's foul-water systems is as a transport medium.

The questions that arise are whether we should integrate human waste treatment into the urban environment to reduce the distance travelled by waste and, if so, how this can be done safely. Codes of practice emphasise the undesirability of having many sewage treatment works in a limited area (11).

Parkland and green landscape are important urban features and one way forward is to stabilise human and other organic wastes to provide valuable nitrogen and organic matter that can be used beneficially in these areas as an effective soil additive, reducing the requirement for other fertilisers.

Aerobic digestion of human waste with a composting toilet (see Figure 9.5) can, given the correct conditions, reduce the volume of waste by 90 per cent (12) through a continual process of solids breakdown and evaporation of moisture. A small extract fan draws air from the room through the pan and compost chamber to the outside via a vent pipe. Designed correctly, such a system can obviate the need for additional mechanical ventilation to the room.

Composting toilets are widely available for single dwellings but might be adapted by increasing the volume of the storage chamber and designing a suitable WC layout to serve a block of flats or offices. Space permitting, the chamber is usually located in a basement or designed as part of the foundations. Organic kitchen wastes may also be introduced into the system via a refuse chute provided from the kitchen.

After a period of 1 to 2 years, the composter is emptied and the compost used in gardens and parks. Depending on the use of additional bulking

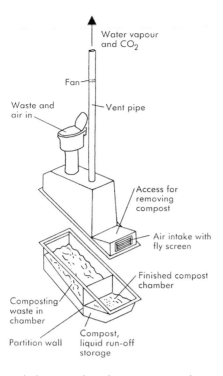

Figure 9.5
Composting toilet

Water vapour and CO$_2$

Fan

Waste and air in

Vent pipe

Access for removing compost

Air intake with fly screen

Finished compost chamber

Composting waste in chamber

Partition wall

Compost, liquid run-off storage

agents such as wood shavings, liquid compost may be removed at more frequent intervals from the chamber and can be spread on the landscape (13). Maintenance and care of composting systems must be understood by the users. For instance, at the Ekoby village in Sweden where a demonstration project was undertaken, many people were unaware that their composting toilets needed emptying (14).

An alternative form of human waste treatment is the septic tank, which is a widely understood technology and has been used successfully in areas without mains sewerage. A septic tank is a type of settlement tank in which sewage is retained for sufficient time for the organic matter to undergo anaerobic digestion. The processes of solids separation and digestion produce a liquid which requires secondary treatment to produce a high quality effluent that can then be discharged safely to the environment. The sludge collected in the bottom of the septic tank will require emptying from time to time. In Europe, between one- and two-thirds of all sludge is spread on farmland (15). Other common methods of dealing with sludge can be found in Reference 16.

Reed beds

Secondary treatment of septic tank effluent can be performed by reed beds instead of, more typically, a below-ground leech field. These are self-contained wetland ecosystems in which complex soil-based microbiological processes promote the degradation of organic and chemical materials and low concentrations of dissolved metals. The reeds themselves are necessary for a variety of reasons, including to introduce air via their roots and therefore promote aerobic digestion. The waste water is delivered either over the surface of the reed bed, and then flows downward (vertical flow), or via a feeder trench at the front of the bed, and then flows horizontally (horizontal flow) as illustrated in Figure 9.6(a) and (b) (17). For more information on the operation of reed beds, see Reference 17.

Reed beds can be situated locally for a single dwelling or at a central facility for a site that serves, say, 100 people. The application determines the scale, and the appropriateness depends to some extent on the context of the site. Reed beds require open land area, which is typically at a premium in urban areas. However, the integration of a reed bed into urban

Figure 9.6
Vertical (a) and horizontal (b) reed-bed
waste-treatment systems

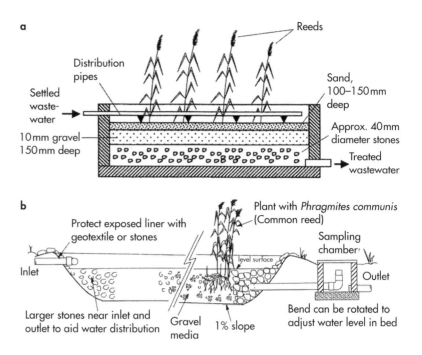

parkland areas could be explored. Approximately 1–2 m² of reed bed area per person equivalent is required for correct operation (19). Important factors in the choice of reed bed size and type are flow rate, organic loading and the required quality of treated effluent.

The topography of the site will also determine the type of reed bed to be used. Vertical reed beds require a fall along their length, and these are best sited on sloping ground to avoid pumping. A final discharge for the odourless, high-quality effluent is also required, and this could be a willow patch, a canal, a lake, or even a water feature.

In practice, in urban situations, finding space for reed beds is not easy. For example, at the Coopers Road development (Chapter 11), which has 664 anticipated occupants on 1.69 hectares of land, providing 2 m² of reed bed area per person would require approximately 8 per cent of the total site area to be devoted to reed beds. Such an area could only be found with great difficulty and so this example illustrates the problem with such an approach in some existing urban areas – but that is not to say that it is impossible.

9.4 Sustainable urban drainage

The integration of urban surface water retention and drainage in the context of the landscape has been discussed in Chapter 4.

A complementary strategy could be to minimise our impact on the natural hydrology of the region. Work carried out as part of the Local Agenda 21 Initiative (20) and illustrated in Figure 9.7 shows how the development of roofs, roads and other impermeable surfaces prevents the natural absorption of rainwater into the ground. Heavy rainfall quickly produces large volumes of water that must be carried away to drain and in the process pollutants are swept from these surfaces into the drains also. A summary of water quality in storm water run-off in urban catchments can be found in Reference 21.

Allowing rainwater to permeate brownfield and potentially contaminated sites could lead to groundwater contamination (22). Strategies including impermeable membranes laid below a permeable surface can retain the benefits of storm attenuation while allowing the rainwater to be collected and then suitably treated. It has also been suggested that natural biological processes within such systems can break down some low-level hydrocarbon pollution (23).

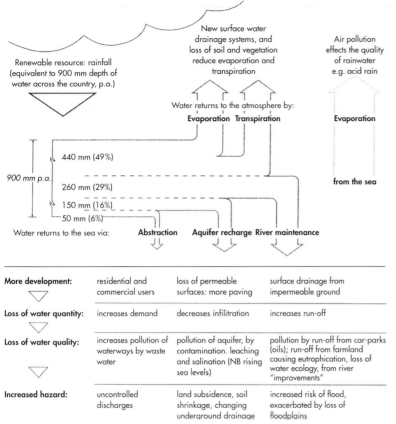

Figure 9.7
The effects of development on the
existing hydrological cycle (UK)

More development: ▽	residential and commercial users	loss of permeable surfaces: more paving	surface drainage from impermeable ground
Loss of water quantity: ▽	increases demand	decreases infiltration	increases run-off
Loss of water quality: ▽	increases pollution of waterways by waste water	pollution of aquifer, by contamination. leaching and salination (NB rising sea levels)	pollution by run-off from car-parks (oils); run-off from farmland causing eutrophication, loss of water ecology, from river "improvements"
Increased hazard:	uncontrolled discharges	land subsidence, soil shrinkage, changing underground drainage	increased risk of flood, exacerbated by loss of floodplains

An example of soakaways as a means of sustainable urban drainage is examined in Chapter 13.

9.5   Waste water strategies in practice

The Cambridge Botanic Garden Education and Interpretation Centre, a proposal for a new building for the University of Cambridge Botanic Garden, demonstrates a water use and waste treatment strategy at scheme-design stage. The site is located in an urban environment, 1.5 km from Cambridge city centre, and is currently partly occupied by a bowls clubhouse and bowling green.

The analysis involves minimising water use and maximising the ability of the site to provide the predicted water requirements of the building and the landscape. Water is available from collecting rainfall and from the underlying chalk aquifer via a borehole. The availability and capacity of both supplies will depend on location. On this site, for example, the average annual rainfall over the past 30 years was 557 mm, or about 1.5 mm per day (24). The Botanic Garden has an existing borehole into the aquifer but has an extraction licence for irrigation purposes only.

In order to comply with current statutory regulations (25), it will be necessary to provide mains water for drinking, catering and washing. Low-water-use appliances will reduce the amount of mains water used for washing.

There will be a constant and predictable water demand and waste loading due to the office workers in the building. Garden visitor numbers, however, can vary from none to 3,700 per day (26) on special occasions.

Waste systems relying on local bacteriological treatment or storage volumes are very sensitive to large variations in waste flow rates. Septic tanks with reed beds work best with a fairly constant throughput, more typical of a domestic or office situation.

In this particular case, trying to provide composting toilets to meet the peak load is impractical because of the storage volumes required. Equally, the site area would be insufficient to treat this load with reed beds.

One strategy is to treat the office waste locally, either with composting toilets or a septic tank and reed bed, but to divert the visitors' waste to the foul sewer.

Figure 9.8 illustrates the proposed water and waste schematic.

Figure 9.8
University of Cambridge Botanic Garden water and waste schematic

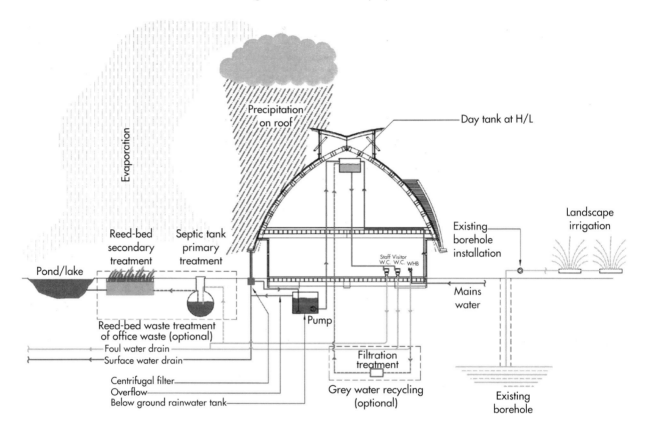

Rainwater incident on the roof will be collected and stored in a below ground tank. The depth of the tank below ground will keep the stored water in darkness and at 10–20°C during most external conditions to minimise bacteria multiplication. Rainwater collects atmospheric pollution and debris from roofs and can become faecally contaminated from bird droppings. The first flush from the roof contains the highest contamination and new filtration systems (27) divert most of this water to drain before collection begins. The stored water is relatively clean but not potable. It may therefore be used for flushing the visitors' WCs and for irrigation and window cleaning.

Some concerns have been raised about the safety of using rainwater, which may be contaminated, for WC flushing (28). As a result of these concerns, a chlorinated day-storage tank will be provided to supply the flushing WCs and will also be a source of mains water backup if necessary.

The remaining site area consists of parking, paving and soft landscaping. Early indications show that the land is not contaminated and therefore porous paving systems and soft landscaping will be used over the remaining site area, allowing rainfall to permeate into the ground.

9.6 Conclusion

It is clear that waste reduction and recycling are key elements in sustainable design. At the building level, a waste strategy should extend beyond the site boundary and consider the opportunity waste offers us for, say, energy generation, or perhaps simply as a rich compost. At the city level, designers and planners are beginning to take the view that an integrated waste strategy, such as that developed in Malmö, encourages beneficial relationships between waste producers and those who can put that waste to good use. Ultimately, the options for waste treatment are similar to those of energy: scale and context will define a variety of appropriate solutions.

9.7 Guidelines

1. Minimise demand to minimise waste.
2. Use waste as a material resource or an energy resource.
3. Minimise water use in transporting and treating human waste.
4. Use rainfall incident on the site.

University of Cambridge Botanic Garden Education and Interpretation Centre: Project Principals

Client	University of Cambridge Estate Management and Building Service
Users	University of Cambridge Botanic Garden
Architects	Edward Cullinan Architects
Services Engineers	Max Fordham LLP
Structural Engineers	Buro Happold
Quantity Surveyors	Gardiner & Theobald

REFERENCES

1. Anon. (2000), Waste Strategy 2000 England and Wales. Department of the Environment, Transport and the Regions, HMSO, Norwich.

2. Jones, P. (2001), What next for waste? Energy and Environmental Management, Nov/Dec 2001, pp. 10–11.

3. As reference 1.

4. As reference 1.

5. Anon. (2001), Municipal Waste Management Survey 1999/2000. Department for Environment, Food and Rural Affairs, HMSO, Norwich.

6. Anon. (1999), Projections of Households in England 2021. Department of the Environment, Transport and the Regions, HMSO, Norwich.

7. Anon. (1999), Ville et Écologie. Ministère de l'Équipement, des Transports et du Logement, Ministère de l'Aménagement du Territoire et de l'Environnement, Paris. pp. 65–66.

8. Anon. (2001), Liquid Asset: Recent Developments in Water Saving. Ecotech (4) November, pp. 34–38.

9. Houghton, J. (Chair), (1993), Cm 2181, Royal Commission on Environmental Pollution, Seventeenth Report, Incineration of waste. HMSO, London, pp. 39–41.

10. MacKenzie, D. (1998), Waste not. New Scientist, 29/8, pp. 26–28.

11. BS 6297:1983: Design and Installation of Small Sewage Treatment Works and Cesspools. British Standards Institution, London.

12. www.clivusmultrum.com

13. Bunn, R. (1994), Living on Auto. Building Services Journal (7), pp. 22–24.

14. Mills, T. (1994), Sufficiently Swedish. Building Services Journal (7), pp. 30–31.

15. As Reference 10.

16. Pickford, J. (1997), Technical Brief No. 54: Emptying Latrine Pits. Waterlines (16) 2, pp. 15–18.

17. Griggs, J. and Grant, N. (2000), Reed beds: application and specification. Good Building Guide 42, Part 1. Building Research Establishment, Garston, UK.

18. Cooper, P. F. et al. (1996), Reed Beds and Constructed Wetlands for Wastewater Treatment. WRC plc.

19. Centre for Alternative Technology (1998), Private communication.

20. Barton, H. et al. (1995), Sustainable Settlements: A Guide for Planners, Designers and Developers. University of the West of England and The Local Government Management Board.

21. Anon. (1983), Design and Analysis of Urban Storm Drainage. The Wallingford Procedure, Vol. 1, Principles Methods and Practice. Wallingford, UK.

22. As Reference 20.

23. Anon. (2000), Storm water source control system 2000. Formpave Ltd, Coleford, UK.

24. University of Cambridge Botanic Garden (2002). Private communication.

25. Anon. (1999), The Water Supply (Water Fittings) Regulations 1999. Department for the Environment, Food and Rural Affairs, HMSO, Norwich.

26. University of Cambridge Botanic Garden (2002). Private communication.

27. Sayers, D. (1999), Rainwater Recycling in Germany. Proceedings from a visit to WISY AG, Frankfurt, 17–19 August 1998. National Water Demand Management Centre, Environment Agency, Worthing, UK.

28. Grey, R. and Mustow, S. (1997), Greywater and Rainwater Systems – Recommended UK Requirements, Final Report, 13034/2. Building Services Research and Information Association, Bracknell, UK.

10

Summary

Randall Thomas

To be successful our cities will need to address all aspects of sustainability. Getting the "structural" issues right is a necessary but not sufficient criterion.

The key environmental issues can be visualised as a triangle of form/density, movement/transport and buildings (energy consumption/production). More compact cities are likely to require less infrastructure and should need less energy for transport. The energy requirements for the buildings may be higher, though, and higher densities will mean it is more difficult to use solar energy. This may be compensated for, in part, by the possibility of distributing the energy collected (or recycled in the case of wastes) more efficiently. Energy supply will be a volatile field but, with a bit of luck, we shall soon see the first major steps towards a society based on renewable energies.

Issues that are more urban than building-related include transportation, large-scale form and density, landscape and biodiversity, collective systems of energy supply and distribution, and water and waste treatment. We can expect some major changes in policies. Movement and transport are likely to shift away from cars and towards walkable communities, bicycles and mass transport. Energy and pollution are not the only driving forces for this shift, which will favour increased densities. Reducing congestion is another important factor.

Landscape will play a vital role in sustainable cities. The "space between the buildings" will, among many other advantages, nourish our aesthetic sense, improve the air we breathe and help save us from the deluge. The buildings and landscape will work together. A major shift in the vision of sustainable cities is the potential of buildings to produce as well as consume energy.

Sustainable cities will require fewer resources and will carefully select the ones they do need, taking into account the effects over the lifetime of their use.

We are in a period of transition. Very few urban projects currently address all of the relevant issues. Our knowledge of the possibilities and the interactions is only gradually increasing. Similarly, the idea of professionals such as local authority planners, designers and engineers all working together is only slowly gaining acceptance – and the concept that they would all collaborate with the users of the urban environment and the occupants of its buildings is novel. Success will require a great deal of effort at ground level. The days when an architect like Le Corbusier could design a city from an aeroplane are probably over.

The case studies that follow illustrate how the general principles discussed in the first part of the book can be applied. They have been written by people who know that design matters and that it can help create cities of delight. The studies mainly deal with the environmental aspects of sustainability but develop other themes as appropriate. Each has its own emphasis and approach and in this reflects the variety of the city itself. The discussion of

Coin Street deals with user participation in the design and management process. BEDZED covers a host of social and environmental issues including mixed use in a very dense setting. Coopers Road is a study of the regeneration of a community and Malmö looks at how to recreate a large part of a city of architectural and biological quality. The importance of imaginative design in creating environmentally responsible and exciting buildings is shown by the study of, for example, the Contact Theatre. How the urban space can be sculpted to respond to the sun is indicated in the Parkmount study. Often one is struck by how local opportunities have been seized upon to strengthen the urban fabric.

There is a world of detail on the road to environmentally friendly cities and the study of CASPAR, with its examination of what it really takes to deal with urban noise and simultaneously reduce energy consumption, gives a flavour of this.

The transition to sustainable, carbon-neutral cities will not be easy and will undoubtedly involve developing the city and the area outside its walls together. There won't be one unique solution. Instead, in a way that is similar to the output of photovoltaic panels, we are likely to see many solutions that give us 95 per cent of optimal performance. If this proves to be true, it will be tremendously encouraging because it will favour diversity, a sense of place and a feeling for the particular. Taken together, the case studies that follow, drawn from UK and European experience, address the main environmental issues of sustainable urban design. They show how we can start to create cities that once again are humane.

PART TWO

11

Coopers Road Estate regeneration: Southwark, London

David Turrent

11.1 Background

The Peabody Trust are actively involved in a number of urban regeneration initiatives across London. Their approach is a holistic one, embracing social, economic and community development issues as well as physical improvements. In 1999 Peabody joined forces with Southwark Housing and appointed ECD Architects to prepare a masterplan for the redevelopment of the Coopers Road Estate. Shortly afterwards they acquired an adjoining site, fronting the Old Kent Road, known as "Success House" and invited proposals from ECD for a mixed-use high-rise building. Concurrent with these activities, Peabody have been carrying out a "community mapping" exercise, liaising with a developer to redevelop the site of a neighbouring redundant pub and planning improvements to Kent House, an existing Peabody estate that abuts both sites. The redevelopment of these sites will act as an important catalyst for regeneration of the wider area. Planning approval for the Coopers Road site was obtained in November 2001 and a start on site is programmed for January 2003. It is anticipated that a planning application for Success House will be made in early 2003.

11.2 Site context

The Coopers Road Estate was built in the 1960s and occupies a 1.69 hectare site north of the Old Kent Road (see Figure 11.1). It consists of 196 dwellings in five blocks, varying in height from three to eleven storeys. Each block is served by a central stair with deck-access to front doors. The existing density is 358 habitable rooms per hectare.

The surrounding area is dominated largely by council housing estates, with the London Borough of Southwark Astley Cooper estate to the west and the Corporation of London's Avondale Estate to the east, with its three distinctive high-rise towers. The Old Kent Road itself is a busy thoroughfare, effectively the main road into the capital from Europe, bounded in the main by an assortment of undistinguished retail "sheds".

The existing open spaces on the Coopers Road Estate consist mainly of grassed areas with a few trees. There is no sense of ownership of these spaces and they provide minimum amenity value. The play areas are not well maintained and have limited equipment. Generally, the open spaces are intimidating and underused.

Success House occupies the site of 419–423 Old Kent Road, adjacent to two existing Peabody buildings, Kent House South and Kent House North. The existing building, which was previously used as office and storage space, is three storeys high with a footprint of 600 m².

Both sites are about 15 minutes' walk (approximately 1.3 km) from Bermondsey station on the Jubilee Line and are well served by bus routes to the Elephant and Castle, the West End and St Paul's. They are also close to

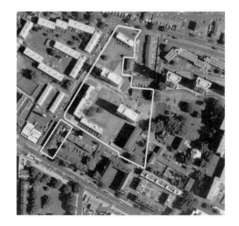

Figure 11.1
Aerial view of the existing site

the Bermondsey Spa Regeneration Area and Tower Bridge, giving links to the City. Local amenities include shopping in the Old Kent Road, a Tesco supermarket and Burgess Park.

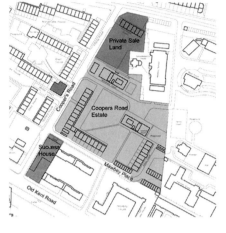

Figure 11.2
Site plots

11.3 Project brief

The Coopers Road Estate is in need of considerable investment and its 1960s design is increasingly unpopular and unsuitable. The communal spaces are abused by sections of the community; the rubbish chutes are blocked with furniture and are considered fire hazards, and the open spaces have poor visual surveillance. In 1999 Southwark Housing made the radical decision to demolish the estate and redevelop it. In doing so they have formed a partnership with the Peabody Trust, who will acquire the majority of the new housing built and manage the whole estate.

The brief for Coopers Road was for a mixed-tenure development of 154 new homes for rent and 36 flats for sale, the latter to be built on a separate plot at the northern end of the site (see Figure 11.2). All homes for rent are to be to Lifetime Homes Standards (1), with some provision for full wheelchair accessibility. The scheme is to be built in three phases to allow the gradual de-canting and demolition of existing blocks. The anticipated density of the new development is 355 habitable rooms per hectare.

The brief for the adjacent Success House site is for a mixed-use development with basement car-parking, in a new building, to support the Coopers Road Estate community and aid the regeneration of the local area. The facilities within will include the relocated existing Youth Club, a community café, flexible office accommodation, thirty flats for key workers (2) and thirty-six flats for sale. In addition to the functional brief, Peabody have aspirations for a landmark building that will demonstrate advanced concepts in sustainable design aimed at minimising CO_2 emissions.

Figure 11.3
Consulting the tenants

11.4 Consultation

What the estate does have is a strong sense of community and although only 50 per cent of the tenants have chosen to return, they are an important ingredient for the long-term success of the project. The tenants have formed a Steering Group with representatives from Peabody and Southwark Housing and they have been closely involved in the development of the masterplan through a series of meetings and workshops.

At key stages in the design process larger events were staged to involve all the returning residents. At stage one, tenants were encouraged to record their desires and dislikes on post-it notes stuck to posters in a mobile "office". This wish-list informed the initial masterplan proposals, which were then presented to the estate at a number of fun-days (see Figure 11.3), with questionnaires available for the tenants to record further comments. Neighbouring estates were included in these events to encourage local awareness and good relations.

Areas in which the tenants influenced the design include the choice of heating system for the scheme and the layout of the homes. When given the choice of individual boilers or a central heating system for the whole estate, the tenants favoured a system similar to their existing district heating system, the view being that any faults would be repaired quickly as they affect a number of people, not just an individual. The comments received on the unit plans changed the proposed layout of the two-bedroom flats. It was suggested that the occupants would benefit from separate WCs and bathrooms, and the plans have been revised to suit.

The consultation process will continue through detailed design into the construction phase of the project and will address future management as well as design issues.

11.5 Design strategy

The new social housing planned for the Coopers Road Estate has been designed around four courtyards. The courtyard form, which evolved through consultation with the residents, encourages a sense of community and engenders a strong sense of identity. It also creates a clear hierarchy of private, semi-private and public spaces and provides a good model for urban regeneration (see Figure 11.4).

Each courtyard consists of approximately forty homes in a mixture of one- to three-bedroom apartments and three- or four-bedroom family houses, providing a balanced community within a composition of four-storey flats and three-storey houses (see Figure 11.5). This arrangement has been designed to be flexible to meet future changing needs and developments in living patterns.

The design principles

1. To restore the fabric of the city, with streets, courts, mews and gardens. The form of the courtyard housing continues an urban design theme from the nineteenth century, from when local maps show a pattern of streets with houses grouped together in short terraces (see Figure 11.6).
2. To integrate architecture and landscape to provide attractive, legible and easily maintained private and public spaces. Large- and small-scale vegetation are located in response to the scale of the surrounding architecture, whilst robust materials are combined with generously planted semi-mature trees to create a sense of greenness across the site.
3. To develop a sense of community ownership individual houses and ground-floor flats have gardens to the front and rear, which create well-defined private space. The rear gardens face on to the larger communal garden (see Figure 11.7), which measures 21 × 34 m, comparable in size to a small London square or churchyard.

The design concept for Success House is for a sustainable high-rise building that provides an important visual marker; a gateway building to the City with stunning views across the West End (see Figure 11.8). The base of the building occupies the entire site area and provides five floors of

Figure 11.4
Proposed aerial view of Coopers Road

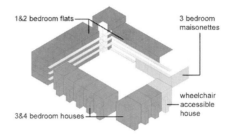

Figure 11.5
Different unit types make up the courtyard community

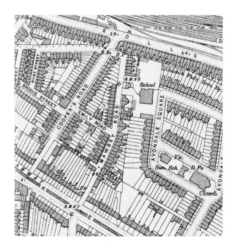

Figure 11.6
The site in 1896

Figure 11.7
Courtyard gardens: private and communal

Figure 11.8
Artist's view of Success House

accommodation over a basement car-park. The remainder of the building is a fifteen-storey tower separated from the base building by a service floor. The tower has a smaller footprint, with three flats per floor served by two lifts, each apartment orientated towards the City-views and away from the noise and pollution of the Old Kent Road.

11.6 Landscape

The aim of the landscape design to the new estate is to create both a physical continuity between the four new courtyard blocks and a highly legible space that is easy to understand. The grain of the new estate allows long views and an ease of orientation that creates a sense of physical security. In addition the landscape scheme aims to achieve a sense of well-being at many levels. It provides the essential contact with the natural world and the changing seasons.

Within each courtyard all the properties at ground-floor level have a small patio garden, which opens into a communal garden for the use of the courtyard residents only (see Figure 11.9). Residents will become involved in the design and management of these spaces, the idea being that they will become a focus for community pride. All of the properties above ground level have generous balconies overlooking the spaces between the courtyards.

Figure 11.9
A detail from the landscaping proposal

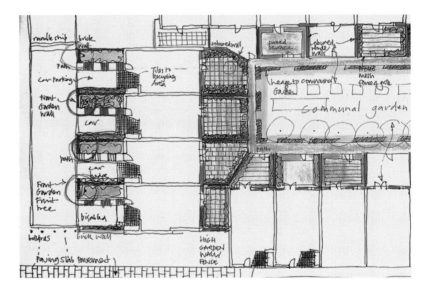

Access roads are designed as "Home Zones" (3) to emphasise the pedestrian's use of the estate roads more than the presence of vehicles. The "Home Zone" roads allow the pedestrians and cyclists to reclaim the streets as they have priority over vehicles that are restricted to just 32 km/h (20 mph). The roads are designed in short runs, intersected by squares and traffic-calming measures. They are envisaged as secure and well-used outside spaces, as the locations for community events.

Success House has been designed to provide sky-gardens for the residents either as green roofs or within the curtilage of their apartments. Apartments within the tower will have access to a roof deck-garden in addition to a private winter garden that serves as a buffer zone between the external and internal environments. The roof of the podium building is envisaged as an extension to Burgess Park, which is located on the other side of the Old Kent Road.

Many of the tenants have lived on the estate or in the neighbourhood for 40 years and are looking forward to having their own garden. The communal gardens are secure spaces that are particular to each courtyard block. Gardening clubs and workshops will be organised to encourage and assist tenants who may never have had a garden before. Within the communal gardens there are places for gardening, cycle storage, composting, young children's play, seating and picnics.

11.7   Sustainability

From the outset both client and design team have shared a commitment to make Coopers Road and Success House a model of sustainable urban regeneration. First and foremost this means designing a scheme that will provide good-quality, appropriate accommodation not just in the short term but 100 years from now.

Orientation and solar access were foremost in the consideration of the planning of the Coopers Road courtyards (see Figure 11.10). The lower three-storey houses are placed to the south of the higher four-storey flats, and roofs are designed to face south wherever possible for the future retrofit of photovoltaic (PV) panels. By maximising the daylight penetration into the homes the demand for artificial heating and lighting is reduced. Success House has been designed to offer a screen of photovoltaic panels to the south that doubles as an acoustic buffer to the traffic along the Old Kent Road.

Flexibility and adaptability have also been key issues for the design team. For example, service risers in Coopers Road are located on the outside of the building for ease of access and in both developments have been oversized to facilitate the installation of future technologies as they become economically viable, such as PV or rainwater recycling.

In addition to this our sustainable strategy is focused on six key areas:

* Energy and CO_2 emissions
* Water conservation
* Materials sourcing
* Waste management
* Transport and car use
* Social well-being

Our aim is to design the buildings in such a manner that it is technically feasible to achieve zero CO_2 emissions by 2020 without major modifications to the fabric, services or infrastructure.

Our first priority is to reduce the demand for energy on site. We then need to select an efficient system for delivering energy, and finally we need to devise a strategy for the gradual phasing in of renewable supplies so that our target of zero CO_2 emissions can be achieved by 2020.

11.7.1 Coopers Road

A. The buildings

I. REDUCTIONS IN CO_2 EMISSIONS

Reducing energy demand is principally achieved by good daylighting, passive solar gain, detailing to reduce air infiltration and high standards of thermal insulation. In Coopers Road we considered very high standards of insulation but these proved prohibitively expensive. Instead we adopted a standard that anticipated the April 2002 revisions to the Building Regulations (wall U-value 0.35 W/m²K, roofs 0.25 W/m²K and glazed

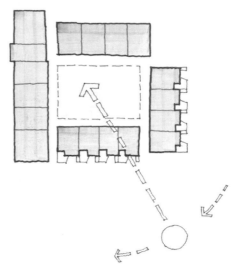

Figure 11.10
Solar access to the courtyards

areas $2.0\,W/m^2K$). With some additional financial assistance from London Electricity residents may also be encouraged to use efficient A-rated appliances.

II. ON-SITE POWER GENERATION

Community heating with combined heat and power has been selected as a suitable way of providing heat and electricity while reducing CO_2 emissions. A detailed feasibility study carried out by Max Fordham LLP concluded that despite the higher initial capital cost, the payback period for the system would be less than 10 years.

One of the main advantages of a central boiler plant is that it allows future changes of fuel supply. A switch to biomass within the next 10 to 15 years could significantly reduce CO_2 emissions; so too could a client/community decision to purchase "green" electricity (see Figure 11.11).

III. WATER CONSERVATION

Water conservation will be achieved by the specification of low-flush WCs and spray taps in the kitchens, along with rainwater-collection systems in which water is filtered and then used for flushing WCs. Rainwater butts will also be provided for some properties.

IV. SELECTION OF MATERIALS

Materials will be selected for their low embodied-energy content and impact on the environment when disposed of. This means preference for timber and masonry rather than plastics and steel. In addition, the contractor will be encouraged to source materials from within a 83 km (50 mile) radius wherever possible in order to minimise CO_2 emissions released in transport. Waste will be minimised by prefabrication of window/cladding elements and standardisation of components.

V. RECYCLING

All dwellings will be provided with a bin within the kitchen units, designed to separate household waste, and recycling facilities will be provided on site.

B. The landscape

I. SOIL REMEDIATION

Site investigation has revealed a low level of site contamination. If tested topsoil is found to be contaminated, the private and communal gardens will be excavated and replaced with neutral imported topsoil. The new 300–500 mm depth of topsoil is sufficient to satisfy the remediation requirements.

II. WATER CONSERVATION

Porous paving has been chosen to allow rainfall to run off and percolate through the subsoil to allow natural drainage. The site is both conventionally and naturally drained. The central courtyard gardens and surrounding porous paving act as natural sumps within the core of the site. About half

Figure 11.11
Selecting sustainable energy sources – photovoltaics

the circulation, such as parking bays, use porous paving, whilst the roads are conventionally drained. Rainwater is collected and stored on site in individual water butts. Residents will be encouraged to recycle and reuse the collected water to maintain the efficiency of the rainwater storage.

III. COMPOSTING AND RECYCLING

Residents will be encouraged to recycle kitchen waste. Kitchen and garden waste can be collected and composted in the communal garden areas. Leaf collection, prunings and grass cuttings from the gardens can also be recycled, composted and returned to the estate gardens as a soil ameliorant.

IV. HABITAT CREATION

A hedge of native species – willows, elder and hawthorn – is proposed adjacent to the eastern and northern boundaries as cover and food for birds and insects. Stands of trees – multi-stem birch and hazel – frame the hedge and generate more cover for the mini-habitat. Standard trees – the native gean (or cherry) – are included within the hedge.

V. GREEN ENVIRONMENT

Trees are planted within the streets and define the edge of circulation spaces, home zones and car-parking. Their location, to the northern side of the blocks, prevents overshadowing and does not limit the passive solar gain to the south-facing elevations. However, the roads and pavements are beneficially shaded during summer months. The street trees have been selected from native species or cultivars of indigenous species – hornbeam, ash, pear and sorbus. Fruit trees and useful plants that are attractive to wildlife are planted within the back gardens and courtyards.

Trees assist the health of the environment, with intake of carbon dioxide, emission of oxygen, filtering of dust and contaminants, and also support a range of wildlife species, particularly birds and insects. The building elevations and garden walls form a protective edge to the streets; these are formed into "green walls". Climbers planted at the base of gable walls act as a deterrent to graffiti.

Gardens are planted with a protective hedge to each boundary wall in front gardens. The hedges can be maintained by infrequent but regular clippings, which are a source of green material for composting. Borders are a mix of hardy, drought-tolerant perennials. Dense ground cover, low-level under windows, combined with a mulch layer, covers the topsoil and helps conserve moisture. Planting to the communal gardens has to be robust to withstand year-round use (4).

VI. SELECTION OF MATERIALS

The inner-city environment of the estate requires a range of robust finishes to surfaces and boundaries, particularly for the public realm, including Tarmac roads, concrete paving slabs and porous concrete blocks. Garden walls in conjunction with the architectural form present a secure perimeter to the blocks. The walls create a protected and beneficial environment for the healthy growth of the evergreen hedges and climbers.

Within the communal gardens the environment is moderated with the use of natural materials. The main components of the garden are timber: seats, bollards, planters, tables, timber decks and edges, all of sustainable heartwood, green oak or douglas fir. All timber is specified from certified and sustainable sources.

Garden boundaries are made from woven-willow panels, supplied from long-established willow beds in Somerset.

The light fittings are selected to minimise light pollution, with downward-facing light sources with reflectors.

11.7.2 Success House

A. The building

A technical strategy has been proposed for the Success House development to reduce the CO_2 emissions to zero by 2020. The base requirement is for a high-performance building designed to require little heating because of its super-insulated roofs, high levels of recycled insulation in the walls and airtight construction to maximise the thermal performance. In addition heat recovery is used on any mechanical extract systems and passive-stack ventilation systems are used to avoid humidity build-up (see Figure 11.12).

Figure 11.12
Success House energy strategy

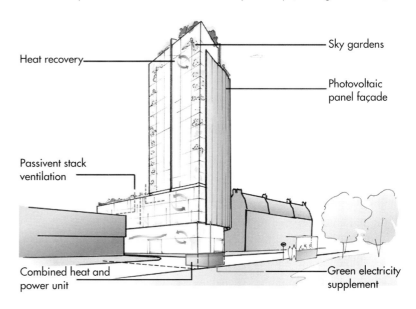

I. REDUCTIONS IN CO_2 EMISSIONS

The CO_2 emissions from energy used to power or heat Success House are reduced by combining electric heating to residential areas with a Combined Heat and Power system to heat the remaining areas of the building, thus maximising the performance of the CHP system. The intention is that the client produces electricity on site by a combination of façade-mounted photovoltaics and the CHP system, topped up by electricity from a Green Electricity Tariff. The CHP system, though initially powered by gas, may run off urban waste or biomass in the future.

II. ON-SITE POWER GENERATION

The façade of amorphous silicon photovoltaic panels serves not only to produce electricity sufficient to power the circulation areas but also screens the building from the noise of the Old Kent Road and expresses the energy-conscious nature of the building.

III. WATER CONSERVATION

Rainwater and grey water will be recycled using a proprietary system of water storage and treatment.

IV. SELECTION OF MATERIALS

The embodied energy within the materials and components used for construction, both in their manufacture and transportation, will be assessed in order to reduce levels wherever possible. The ideal sourcing distance for materials is from within a 83 km (50 mile) radius. The long-term performance of the materials will also be considered in order to maximise the periods between removal and replacement of building elements.

Figure 11.13
Installing prefabricated elements

V. PREFABRICATION

The prefabrication of building elements (see Figure 11.13) will help in many of the objectives outlined above, such as airtightness and CO_2 emissions in transportation (the result of a reduced number of trips to site). Generally, waste is also reduced in factory construction. Additional measures such as on-site waste separation and recycling will be encouraged to minimise construction wastage.

VI. TRANSPORT

The location of the site adjacent to existing public-transport routes within an established urban environment minimises the reliance on car use. Car pooling was also considered as a way to reduce the ownership of cars on site. The basement car-park could easily operate as a secure car pool.

VII. SOCIAL WELL-BEING

Elements of the design discussed above also serve to improve the social environment within the building. The acoustic protection offered by the photovoltaic shield and the peripheral winter gardens modify the internal environment to the benefit of the residents The mixed-use activities within the development ensure that the building is active 24 hours a day, 7 days a week.

B. The landscape

I. REDUCTIONS IN CO_2 EMISSIONS

Sedum planting on the podium building and the tower roof garden will improve the quality of the immediate environment by their intake of carbon dioxide, emission of oxygen and filtering of contaminants in the air.

II. HABITAT CREATION

The relatively large area of sedum plants on the podium-building roof will encourage local wildlife such as birds and butterflies back to the area.

The sedum roof interrupts the rain flow. If the roof were finished in a traditional building material the run-off would pass directly off the building and into the local drainage system. The sedum planted surface retains water for plant growth, lessening the impact of the building on the local infrastructure.

Figure 11.14
Success House model

IV. ENVIRONMENTAL BUFFERS

Within the building, sky-gardens will be built into the private apartments (see Figure 11.14). The intention is for these spaces to act as buffer zones between the consistent environment of the internal spaces and the erratic environment outside. The spaces can be closed or opened to the outside depending on the levels of wind, noise and temperature.

Figure 11.15 demonstrates how the goal of zero CO_2 emissions will be achieved for Success House.

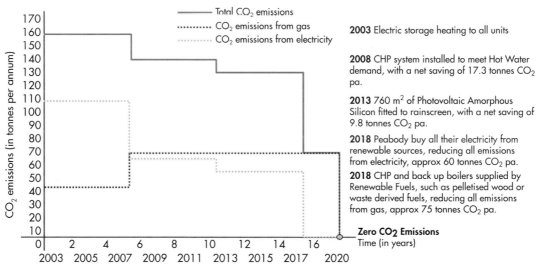

2003 Electric storage heating to all units

2008 CHP system installed to meet Hot Water demand, with a net saving of 17.3 tonnes CO_2 pa.

2013 760 m^2 of Photovoltaic Amorphous Silicon fitted to rainscreen, with a net saving of 9.8 tonnes CO_2 pa.

2018 Peabody buy all their electricity from renewable sources, reducing all emissions from electricity, approx 60 tonnes CO_2 pa.

2018 CHP and back up boilers supplied by Renewable Fuels, such as pelletised wood or waste derived fuels, reducing all emissions from gas, approx 75 tonnes CO_2 pa.

Zero CO2 Emissions

Predicted reduction in CO_2 emissions for Success House to achieve zero CO_2 within a 17 year period

Figure 11.15
Table CO_2 emissions for Success House

11.8   Summary

The aim of the project is to provide a model for sustainable development in an urban context that is affordable and replicable. The buildings are capable of developing over time and the servicing strategy is simple and robust, designed to support change. Key points are:

- mixed-tenure sustainable urban regeneration;
- a model of good community consultation;
- courtyard buildings enclosing communal gardens;
- mixed-use eco-tower;
- community heating and CHP;
- design strategy to achieve zero CO_2 emissions by 2020.

Project principals

ECD Architects
ECD Energy and Environment
BPP Construction Consultants
Price and Myers Structural Engineers
Max Fordham Services Engineers
Jenny Coe Landscape Architects
Philip Pank Partnership Planning Supervisors

REFERENCES

1. Lifetime Homes Standards as defined by "Meeting Part M and designing Lifetime Homes" published by the Joseph Rowntree Foundation, 1999.

2. Key workers include teachers, nurses, cleaners and other low-paid workers vital to the economy of London.

3. A home zone is a street or group of streets designed primarily to meet the interests of pedestrians and cyclists rather than motorists, opening up the street for social use. Legally, neither pedestrians nor vehicles have priority, but the road may be reconfigured to make it more favourable to pedestrians.

4. Particular thanks are due to Jenny Coe for her contribution to the text on landscape design.

Parkmount: streetscape and solar design

Richard Partington

12.1 Background

Northern Ireland, and Belfast particularly, is being viewed with interest as an emerging growth area with potential for urban regeneration. The city is looking to structure development through a series of planning initiatives, development briefs and design guides, but the issues that will shape future growth and the mechanisms for achieving it are fraught. This case study considers a housing project, now under construction, that was planned from the outset to be a model for regeneration and sustainable development.

In Belfast the tentative steps towards a more peaceful and integrated society have yet to be reflected in the physical landscape, which is still marked by the heavily fortified police stations, the provocative murals and painted kerb lines, and the oppressive walls that separate communities (1). Everywhere there are constant reminders that Belfast is a patchwork mosaic of neighbourhoods "differentiated by religion" (2).

These are the characteristics of the southern end of Shore Road and our site, known as Parkmount, which had been earmarked for redevelopment from as far back as 1997. To the south-west is the mainly derelict and notorious Mount Vernon estate, a loyalist and paramilitary stronghold, scheduled for demolition. Immediately to the east is the Shore Crescent estate, a Protestant community planned in the 1970s around a series of "secure" dead-end roads and culs-de-sac, as was the practice for housing layouts of the time, guided in part by Oscar Newman's concepts of "defensible space" (3). Further to the east again is the new motorway, which runs parallel to Shore Road along the edge of Belfast Lough. Figure 12.1 shows a map of the immediate area.

Figure 12.1
Map of the area to the west of Belfast Lough between the Antrim Road and Shore Road

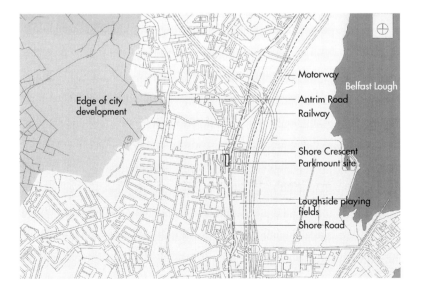

12.2 Brief

In 1997 the Northern Ireland Housing Executive, the main public body responsible for delivering housing, established a project team to promote new ideas in housing design with the intention of building these into a demonstration project. The Parkmount site was identified and the NIHE then looked to private developers, through a tender process, to deliver the project and take the risk for marketing and sales. There was a process of negotiation and discussion with the winning tenderer, the Carvill Group, that clarified at the outset the aims of all the parties and set down the benchmarks for assessing the environmental performance of the scheme.

The accommodation consists of fifty-six two-bedroom apartments (approximately 60 m²) with two smaller one-bedroom apartments. Forty-six of the apartments have wheelchair access and internal dimensions that suit wheelchair manoeuvrability.

The key components of the brief are:

- flexible apartment plans to anticipate changes in work patterns and lifestyles;
- creation of a defined "place" with landscaping and safe play-areas;
- a completely secure development with controlled access;
- a logical sequence for marketing and constructing the scheme in phased stages to limit the financial risk;
- simple and reliable technical solutions that will be economic to run and maintain;
- attainment of the BRE EcoHomes Standard (4);
- good design for maximising solar potential with a high research and innovation component centred around the use of PVs.

12.3 Urban context

The urban characteristics of Belfast are now highly unusual within the British Isles. Whilst many nineteenth-century cities in England expanded through the 1970s to 1990s, with the development of city fringe and the erosion of established centres, Belfast retained much of its compact arrangement of merchant city core and "arterial" routes (Figure 12.2, Llewelyn-Davies' "starfish" diagram, shows a clear graphic interpretation of the structure of the city). Shore Road is one such arterial route. The road carriageway is wide and most of the house frontages are set a long way back from the

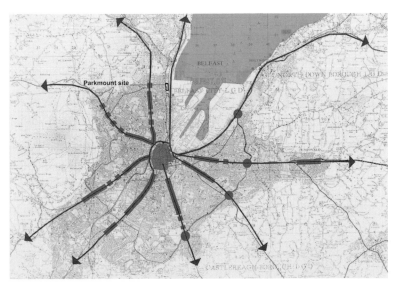

Figure 12.2
The "starfish" diagram illustrates a pattern of linear arterial routes radiating outwards from the city centre

Figure 12.3
View looking south

Figure 12.4
View looking south-west

pavement edge, making it feel exposed and windswept. Most of the buildings are two storeys high and there are extended gaps in the street frontage so there is no sense of "enclosure" compared with a traditional street. However, there is a thriving evangelical chapel, a community centre for the elderly and well-used sports facilities and playing fields, though these have inadequate changing facilities.

The site itself is a long, thin strip of derelict land, aligned on a north–south axis, 3.5 km from the centre of Belfast. A flat area in the centre approximately 35 metres wide and 160 metres long provides the only useful space to build upon. The designers were asked to accommodate more than fifty two-bedroom apartments on the site. The final scheme provides for fifty-six two-bedroom apartments and two one-bedroom apartments, giving a density of 97 units per hectare (290 habitable rooms per hectare). Figures 12.3 and 12.4 show the existing site.

Along its eastern edge, where a row of semi-detached houses previously stood, there is a continuous frontage to Shore Road. To the west the land rises very steeply in a densely wooded escarpment. From the top, nearly 8 metres above the road, there are dramatic views of Belfast Lough; the Harland and Wolf dockyards with their twin yellow cranes; and the city centre. The woodland and western escarpment form a verdant backdrop to the site and beyond this on the far horizon is the dramatic profile of Cave Hill. Figure 12.5 shows the context of the site.

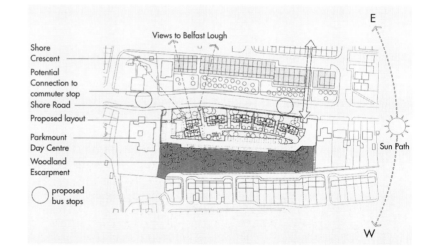

Figure 12.5
Site plan

Working with urban designers Llewelyn-Davies, our intention was to resurrect the concept of a "linear community" with shops, health, leisure and employment along the road and intensification of Shore Road as a public transport corridor (5). The scheme aims to be a beacon, to register sufficient presence and scale on the derelict landscape to become a recognisable landmark and to represent the new optimism and ambition of the city. We hope that on a clear day it might even be visible from the more affluent areas on the south-eastern side of the Lough.

The project was considered as part of a wider study area for which several strategic plans were produced. These plans proposed a better use of the existing infrastructure by adding a commuter stop to the railway on the other side of the road from the Parkmount project. Multiple crossing points would then be provided to give access from both sides of the road and introduce some much-needed traffic calming. New pedestrian routes to the commuter stop would stimulate improvements to frontages and the public domain and make connections through the dead-end/culs-de sac layout of Shore Crescent. The commuter stop was imagined as one of a series of positive interventions, including new changing facilities for the playing fields, a sports hall and clinic or surgery. Each of these would be coupled with existing buildings or uses to concentrate activity and make efficient use of resources. The commuter stop would therefore have a parking area that could be coupled with the sports centre and a "park and ride" scheme to give efficient use at all times and avoid duplication of provision for each of the different buildings. Each new use would also have an associated bus stop.

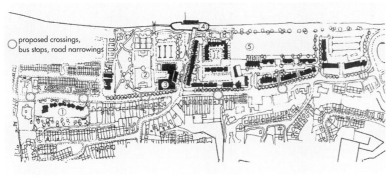

Figure 12.6
Sketch masterplan for Shore Road – proposed building in bold

1. **Parkmount**
2. **Playing field**
3. **Leisure centre**
4. **Commuter stop**
5. **Employment uses**

12.4   Urban design and planning

On the housing site itself our intention is to present a positive edge to the public domain – a proper street frontage. We contended that the scale of the development should relate to the scale and importance of the road itself rather than be determined by the two-storey fragments around the site.

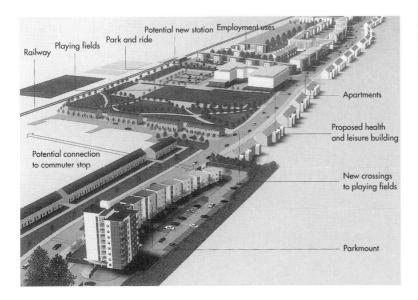

Figure 12.7
Computed-generated view of Shore Road – development masterplan looking south

This issue was considered simultaneously with the main ambition to achieve an exemplary energy performance and to realise the maximum solar potential of the site, considering daylighting, passive solar gain and the use of PVs. We were also asked to create a high-quality landscaped space that would be central to the whole project. This space had to be useable, that is, overlooked, secure, well lit, sheltered and densely planted. The planting was essential to improve the ecological value of the site and to structure the courtyard space and screen the residents' parking.

127

12.5 Site layout and enclosed space

Planning policy in Northern Ireland has recently been revised to reflect the Government's commitment to sustainable development and design quality for residential projects (6). As with other initiatives there is a clear emphasis on creating movement patterns that support walking and cycling and a requirement for each residential development to provide public and private open space and integrated landscaped areas. One of the perceived shortcomings of post-war housing design is the failure to address the issues of ownership and demarcation of space. Landscape areas and open space whose ownership is common or ambiguous is often poorly maintained or vandalised, which in turn has implications for crime prevention and personal security (7). However, in Belfast this is not just a question of providing natural surveillance, and distinguishing "front and back" or "public and private". Understandably, house purchasers expect high levels of visible security and active surveillance and gated developments with security cameras are not uncommon. We were uncomfortable with the idea that Parkmount might be conceived of as a secure "enclave" – this seemed to contradict the notions of permeability and pedestrian movement that underpin regeneration planning.

At Parkmount a number of devices are used to establish clear but secure entrances to the project and to delineate the transition from street pavement to dwelling entrance. In Victorian and Edwardian housing this transition would have been marked by a series of thresholds, each reinforcing the demarcation of private ownership: the gateway, the front garden, the steps up to the front door, portico or porch, etc. At Parkmount, the single shared space has one secure access point, so it is not a public space but gaps between the buildings are placed to give glimpsed views into the courtyard and these are emphasised on the street edge where there are breaks in the low wall and railings. At these points paired free-standing brick pillars with small canopies and lighting suggest gateways and entrances. In the future, with the emergence of a safer and less-troubled society, it will be possible to connect these gateways directly to the landscaped court with footpaths.

The entrances to the low-rise apartment buildings are placed on the protected garden side, so that the garden play-area and parking benefit from natural surveillance. The tower, however, is approached both from the courtyard and via a gatehouse that was designed as an entrance building, incorporating secure cycle storage and possible accommodation for a concierge, on the street edge. Each apartment has full-height glazing and a generous balcony to the living space overlooking the street. The access stairs are placed behind continuous vertical walls of glazing, increasing the sense of transparency and visibility and helping to represent the interior life of the building on the "public" side. Although entrances are on the private-courtyard side, the scheme's frontage addresses the public street. The façades are sufficiently animated and appropriately urbane, that is to say simultaneously capable of being "read" as housing and as part of the scale and structure of the city.

12.6 Solar design

The buildings are organised in a gently sweeping arc aligned with the north–south orientation of Shore Road. We discovered, through a series of massing and site studies, that the question of overshadowing and the possible forms for the roofscape would strongly influence the proposal. The development of the arrangement and massing of the buildings is discussed in the Parkmount case study in *Photovoltaics and Architecture* (8). To maximise the surface areas of the roofs that could present an optimal inclination and orientation to the southern sky (within plus or minus 20° of due south) we decided to concentrate on roof forms that would avoid overshadowing. The form of the lower buildings responds directly to those conditions, with the lowest point of the roof line at the southern end of the site ascending at a constant 5° angle towards the north and culminating in a nine-storey tower. Figure 12.8 shows the final layout of buildings.

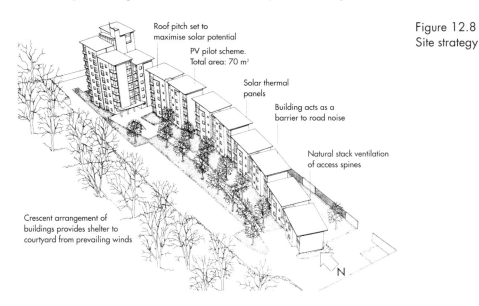

Roof pitch set to maximise solar potential

PV pilot scheme. Total area: 70 m²

Solar thermal panels

Building acts as a barrier to road noise

Natural stack ventilation of access spines

Crescent arrangement of buildings provides shelter to courtyard from prevailing winds

N

Figure 12.8
Site strategy

The roof profile is the distinguishing feature of the project. Compositionally it helps to unify the whole scheme so that each of the repeated blocks is seen as part of a considered whole. The strength and clarity of the roof form is also a clear architectural representation of the central theme of solar design, which has influenced every aspect of the project. Figure 12.9 shows a computer-generated image of the Shore Road elevation and roof profile.

Figure 12.9
Computer-generated image of Shore Road elevation

As the project has developed we have been able to investigate the possibilities for solar design in an urban form that follows a north–south street pattern. This orientation has generated different internal conditions than might have been expected if the streets had followed the classic east–west alignment for solar design. Even though the living spaces have various orientations, more than 80 per cent of the apartments will have good sunlight penetration and many have dual-aspect living rooms, which are ideal for a normal working day as they give direct sunlight in the morning and the evenings.

Those apartments in the tower that do not have ideal orientation, because of the density and the constrained site, have the compensation of the best views in a north-east direction towards the Lough. We have tried to strike a balance between solar design, the streetscape issues discussed above and the other considerations imposed by the setting and the brief. The quality of daylight in interior spaces and the potential for passive solar design seems to have been ignored in so much recent housing, where small windows are the norm. By paying attention to these issues we believe the quality of the internal spaces and their marketability will be greatly increased. Where cost concerns have influenced the external wall treatment we have altered the specification of wall elements (render in place of brick) in order to maintain the large window areas. Figure 12.10 is a sketch analysis of daylighting for the developing scheme.

Figure 12.10
Analysis of daylighting at sketch design stage

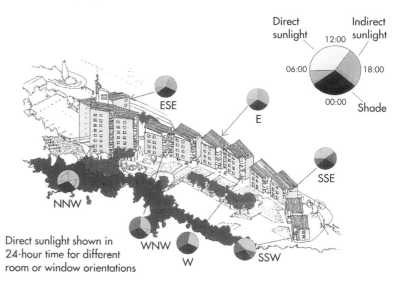

Direct sunlight shown in 24-hour time for different room or window orientations

An array of 70 m² of grid-connected photovoltaic panels is being designed with an estimated annual output of 4,400 kWh/year, which should provide equivalent electricity to meet the annual consumption of two of the apartments. The design for this sample roof area (PV fixing, access, cable routing, inverter design, etc.) can be applied to all the low-rise units in the future. The scheme will use amorphous silicon modules, factory-bonded to the single-ply roofing membrane and protected by a fully weatherproofed transparent polymer coating. The amorphous silicon can be applied in very thin, flexible layers, which makes it highly suitable for use with lightweight roof membranes. On the other half of the same block an installation of solar thermal panels is also proposed, providing pre-heating of hot water to all of the apartments below. One 2.5 × 1.3 m panel, with an effective absorber area of 2.8 m², will be installed per flat. The cost of the revised heating system and the energy saving for each type is going to be monitored.

Although the solar performance could be improved slightly by a steeper roof pitch and a more rigid alignment, we opted for a more balanced solution based on the gentle arc form. The arc is an "invention", an architectural idea, that developed through thinking about several issues simultaneously – an enclosing form, solar design, the street edge, clarity of intention. None of these in itself is a constraint but rather they are all considerations that are assimilated through the design process, leading to something that one hopes will transcend the purely functional or perfunctory.

12.7 Building construction and appearance

The external elevations combine panels of smooth render with textured brickwork. Punched windows in the masonry walls have deep-set window reveals to emphasise their thickness and mass. The additional thickness is a result of the wider than usual cavity, in this case 140 mm, to achieve a significant increase in the U-value (30 per cent better than the Building Regulations requirement). The cavity is filled with a polystyrene bead insulation, chosen for its environmental performance, having a zero ozone-depletion value. The base of each building is emphasised with a full-storey height of blue engineering bricks.

The corners of the buildings have full-height glazed panels with doors opening onto projecting balconies, each with a finely detailed metal balustrade in contrast with the mass of the masonry elements. The clean lines of the corners are repeated around the stair cores and entrances, where transparent glass is used to suggest openness and arrival. Glazing to the stair cores will ensure that distant views can also be enjoyed from here. The glazing, which also contributes to the high thermal performance, will be a timber-framed system incorporating low-E glass to achieve an average U-value of $1.9 W/m^2K$. The wall design and the window specification represent the main departures from standard house-building practice and have raised particular questions on the construction of the tower, for which the effectiveness of the full-filled cavity in protecting the inner leaf from water penetration had to be considered. Structurally the walls are subject to high wind loads at the upper storeys and the relevant safety factors also had to be built in to the design.

The tower presented a difficult challenge for the design team. The developer insisted that the efficiency of the scheme (the ratio of saleable area to total area) should be around 85 per cent. We had imagined that the tower would have the slender proportions of some familiar precedents in London, for instance the well-known Coin Street tower designed by Lifschutz Davidson (9). Whereas the elegant Coin Street tower had less onerous constraints, we needed the service core to access four apartments per floor to achieve the required efficiency, changing the proportion of the tower considerably. Our solution was to treat each quarter of the plan as a vertical element, a tower in itself, attached to the central core and articulated to emphasise its verticality.

Each lower "element" rises to a different height, giving a spiralling effect at the roofline but compromising the suitability of the form for renewable solar energy. This compromise was deemed reasonable to preserve the effect of the tower on the skyline; a bold intervention set against the sombre backdrop of Cave Hill. Figure 12.11 shows a computer-generated image of the tower.

Figure 12.11
Computer-generated image of the tower from Shore Road

12.8   Energy and EcoHomes rating

All of the apartments in the scheme have a predicted SAP (Standard Assessment Procedure) (10) value of more than 96. Seventy per cent of the apartments achieve higher than 100. The solar thermal panels contribute approximately two additional points to the SAP rating for the apartments that they serve. The PV panels improve the score by over 10 points. The apartments with PVs score an average SAP of 114, compared with 105 for the equivalent apartments without PVs, with variations resulting from orientation and vertical position (the top- and ground-floor apartments perform slightly less well).

The SAP rating is very sensitive to the wall construction, as seen above, and the specification of the boiler. All the apartments will have gas-condensing boilers with burner modulation and weather compensation. Replacing this type of condensing boiler with a standard combination boiler would reduce the SAP rating for a standard mid-floor flat from 107 to 95 (11).

The SAP calculation also forms an essential part of the EcoHomes assessment procedure. EcoHomes is an environmental rating that rewards developers (and therefore purchasers) who improve the environmental performance through good design. The EcoHomes scheme provides a credible transparent label for new or converted houses, apartments or shared accommodation.

In the assessment procedure a large number of credits are available for performance in reducing carbon dioxide production resulting from energy consumption. This category also receives a heavy weighting in the final scoring calculation. Parkmount scores well in this area and further improvement could only realistically be achieved by a large-scale installation of renewable energy sources. Figure 12.12 shows a construction detail illustrating how the increased cavity width is achieved.

Typical jamb detail to window (brick wall)

Typical window cill detail (brick wall)

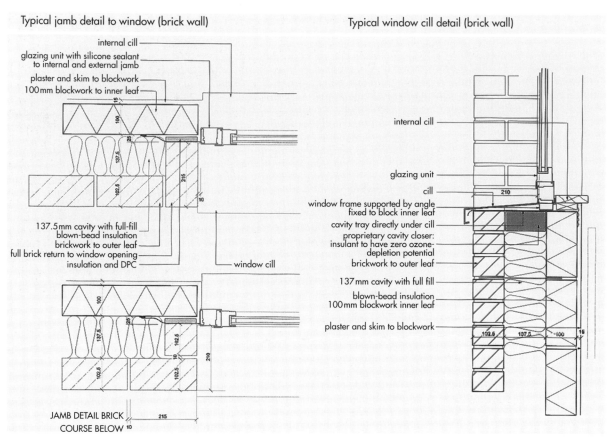

internal cill
glazing unit with silicone sealant to internal and external jamb
plaster and skim to blockwork
100mm blockwork to inner leaf

137.5mm cavity with full-fill blown-bead insulation
brickwork to outer leaf
full brick return to window opening
insulation and DPC

window cill

JAMB DETAIL BRICK
COURSE BELOW

internal cill

glazing unit
cill
window frame supported by angle fixed to block inner leaf
cavity tray directly under cill
proprietary cavity closer: insulant to have zero ozone-depletion potential
brickwork to outer leaf
137mm cavity with full fill
blown-bead insulation
100mm blockwork inner leaf
plaster and skim to blockwork

Figure 12.12
Construction detail of wall cavity

12.9 Internal layout and flexible use

Lifetime Homes (12) are intended to accommodate different uses and access requirements as households meet different needs – raising small children, having visitors to stay or providing for people with mobility difficulties in old age.

The Parkmount scheme is designed around two basic flat types incorporating design features that support the Lifetime Homes principles, with some variation in type: the low-rise types maximise cross-ventilation by having dual aspect living spaces; the high-rise types all have wheelchair access.

Each apartment has a fixed service core to allow efficient distribution of services and to achieve standardisation in the services installation. The remaining area has a load-bearing cross wall and clear span to each half floor to allow the flats to be configured differently, using a demountable system. In the tower apartments structural considerations required all of the cross walls to be load-bearing. The provision of additional telephone points in the second bedroom and living areas will allow for home-working in either of these spaces.

Figure 12.13 shows a typical low-rise dual-aspect apartment plan.

Figure 12.13
Typical apartment plan

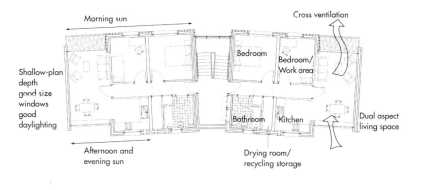

Morning sun

Cross ventilation

Shallow-plan
depth
good size
windows
good
daylighting

Bedroom

Bedroom/
Work area

Bathroom

Kitchen

Dual aspect
living space

Afternoon and
evening sun

Drying room/
recycling storage

12.10 Broader issues: transport, connections, beyond the site boundary

New planning guidance in Northern Ireland has been looking ambivalently at UK mainland practice and responding to pressure on the NIHE to abandon culs-de-sac and introduce networks of connecting streets. A new-housing design guide, *Creating Places*, describes quality aspirations for housing layout based on looped streets. The guide is informed by *Places, Streets and Movement* (13) but, unlike this reference document, is compromised by an adherence to standards, especially with regard to highways design and parking ratios. At the time that negotiations were going on with the planning authority the use of this guide and the quality initiative PPS7, referred to in the section on site layout, had not been fully implemented.

It was envisaged that the Parkmount development would integrate with other regeneration initiatives along Shore Road, including speed-restraint design of the highway itself. The wider environmental design would include good-quality pedestrian crossings, cycle paths and tree planting. However, although the Highways Department encouraged the improvement of public-transport facilities, suggesting a new bus stop closer to the next-door community centre, they were extremely resistant to any suggestions for traffic calming or width reduction of the main road. The Department conceded that the road was unnecessarily wide (most of the through traffic now uses the motorway) but argued that additional capacity had to be maintained as a relief mechanism in the event of an incident blocking the motorway itself. As a result the road will continue to be dangerous for pedestrians to cross and vehicle speeds will remain at unacceptable levels for a residential area. The opportunity to create a new commuter halt on the railway line to the east of Shore Crescent is also unlikely to be realised until Belfast develops an integrated transport plan.

The scheme provides one and a half parking spaces per dwelling, to comply with current Northern Ireland standards, although this is probably more than will be required, based on the car-ownership levels of first-time purchasers. Despite the recent policy reviews, and contrary to mainland practice, the trend in Northern Ireland is to provide for more cars. Because density and parking are directly related this trend is somewhat at odds with the aspiration to realise the full potential of "brownfield" land (14). At Parkmount the high parking requirement will have a direct impact on the quality of the landscaped areas. Even more illogically, the guidance takes little account of the location of projects in relation to parking. We have developed an argument for promoting "linear communities" along transport corridors where there are viable alternatives to car use. Development in any of the areas along the rays in the starfish diagram (Figure 12.2) could have reduced parking provision but the standards only contemplate reduced parking in inner urban areas. The guidance ignores the strong links between parking supply, housing density, location and car use and their impact on the quality of residential environments – the broader view encouraged by the Urban Task Force (15).

The transport and parking issues illustrate the problems that occur when planning policy continues to depend on a standards-led rather than a design-led approach. *Shaping our Future* (16) states that "in order to achieve quality housing schemes, a design led approach will be encouraged. The emphasis will be on flexibility, within reasonable limits, in relation to density, car-parking and road space". For this framework to be effective planning authorities must be encouraged to exercise judgement and discretion informed by clear guidance rather than inflexible standards.

Figure 12.14
Computer-generated image of the tower

12.11 Conclusion

Parkmount attempts to address a range of urban design issues posed both by the specific brief and the current debate on urban design. As the design progressed, the main considerations began to influence our own understanding of what "sustainability" and environmental responsibility might mean for developer-led housing. Below are the key tenets of sustainable urban design that have emerged.

First, the scheme takes a "position" in relation to a broader view of the public realm – to Belfast City, its location, its heritage, the typology and forms of its buildings and streets, and the richness and diversity of its culture. At the same time it attempts to realise the environmental potential of the site.

Second, the scheme acknowledges the immediate requirements and ambitions of its residents. There is a human component – we are designing for a culture in which the private domain is increasingly more central to our existence than a wider sense of society. More and more of what used to be viewed as public life is now lived out in the comfort of the living-room. There is a potential conflict between individual liberties – the right to privacy, for instance – and public aspiration, the desire to use scarce land efficiently, that has to be balanced carefully.

Third, a sustainable neighbourhood has to be successful and desirable. Housing designers have inherited a legacy of social experimentation and unsustainable optimism, characterised by failed housing estates and new-town plans. Now planning authorities and planning guidance has retreated towards the opposite extreme of conservatism and historical sentimentality but we believe there is a demand for a humane modern aesthetic (17).

In Belfast we have been considering diverse and seemingly unrelated issues – the question of security and how it will have an effect on the desirability of the area and the scheme, alongside the possibility of using photovoltaic technology, a technical field that is still in its infancy. In both cases we have tried to look forward; in the case of security, to a time when the scheme may become more permeable and accessible and less inclusive. The buildings have been designed to sustain the changes that may arise from this culture shift. In the case of solar design, we have made the roof planes and surfaces so that they will, when appropriate, provide the optimum orientation for solar panels, anticipating a day when their pay-back period and the universality of their construction techniques make them suitable for every new building and refurbishment project.

Project principals

Client	The Carvill Group
Architect	Richard Partington Architects
Urban Designer	Llewelyn-Davies
Services Engineer	Max Fordham LLP
Structural Engineer	Fergus Gilligan and Partners
Quantity Surveyor	The Carvill Group
Contractor	The Carvill Group

REFERENCES AND NOTES

1. Peace lines, as they are euphemistically known, have been built on the edges of the polarising sections of Belfast's northern and western neighbourhoods and extend in sections up to 1 km long. Though started as impromptu barricades, they now form permanent walls demarcating the sectarian boundaries that have been drawn with

increasingly harder lines since the escalation of "the troubles" in 1969.

2.   McHale, S. (1999), "Terror, Territory and the Titanic", Cambridge Architecture Journal, Scroope 11. The patchwork of neighbourhoods has created urban characteristics that are highly unusual within the British Isles, explaining in part the Province's ambivalent attitude towards UK planning policy and guidance. The Belfast Urban Area Plan 2001 describes the city as "neutral territory" and proposes a commitment to only 50 per cent of development on brownfield land. In Belfast there are areas of undeveloped land which act as buffers between the communities and so there is question as to whether this blighted land, often in the shadow of the peace-line walls, can be included within a definition of brownfield land.

3.   Newman, S. (1973), Defensible Space: People and Design in the Violent City, Architectural Press, London, and Coleman, A. (1985). "Utopia on Trial – Vision and Reality in Planned Housing", Hilary Shipman, London, sought to establish causal relationships between the organisation of buildings and spaces and their potential for natural surveillance. All of these concepts take on new significance in a divided city like Belfast.

4.   Building Research Establishment (2000), Construction Research Communications Ltd, London.

5.   Refer to "Area Development Plans – Sustainable Cities Programme", produced by Napier University, which highlights the positive impact of public transport corridors, interchanges and railheads on overall energy efficiency and urban sustainability.

6.   DOE (2001), Planning Policy Statement 7 (PPS 7). Quality Residential Environments, Northern Ireland.

7.   Alice Coleman's research team attempted to identify "design disadvantagement factors" on problematic housing estates. "Utopia on Trial – Vision and Reality in Planned Housing" (as reference 3).

8.   Thomas, R. (2001), Photovoltaics and Architecture. Spon Press, London.

9.   Coin Street Broadwall (1994), Architecture Today, Issue 52.

10.   The SAP rating system is a method of predicting total energy requirements for dwellings irrespective of location or usage; the end result is a normalised energy rating for comparison purposes. The method is based upon the BREDAM energy model produced in 1985. The method produces a rating on a scale of 1 to 100 for Building Regulations purposes, although in reality values of over 100 can be achieved.

11.   SAP analysis and calculations provided by MFLLP, May 2001, based on the construction issue drawings and specification.

12.   Brewerton, J. and Darton, D. (1997), Designing Lifetime Homes. The Joseph Rowntree Foundation, York.

13.   DETR (1998), Places, Streets and Movement: a companion guide to Design Bulletin 32.

14.   The Belfast Urban Area Plan (2001) describes the city as "neutral territory" and proposes a commitment to 50 per cent development on brownfield land.

15. Urban Task Force (1999), "Towards an Urban Renaissance". DETR, London

16. DOE (2001), SPG 6 Regional Planning Strategic Framework for Northern Ireland, Northern Ireland.

17. Poundbury and Knottley Green, the exemplars of housing and planning guides up and down the country, supposedly hail a new era, but they are in the image of each and every historical period (except our own), all thrown together to form incongruous housing "theme parks".

13

CASPAR II: innovation in housing

Tristan Couch, Max Fordham, David Levitt and Mark Lewis

13.1 Introduction

CASPAR is an acronym of City-centre Apartments for Single People at Affordable Rents. CASPAR II in Leeds is the second development of its type instigated by the Joseph Rowntree Foundation. The changing expectations in the way we live have identified a shortfall in the availability of single-person accommodation. The social and environmental advantages of inner-city living, together with a need for diversity in the way we use these areas, was the inspiration for the CASPAR projects. The development of CASPAR II includes many new and innovative approaches in the building industry, including heat-reclaim ventilation, partnering, prefabricated-pod construction and sustainable urban drainage.

The site is on the edge of the central business and shopping district in Leeds. The main pedestrian mall is a 5-minute walk away, providing easy access to shops, entertainment and work. The local bus network serves the site and the main national train station is a 10-minute walk away. Despite the proximity to the heart of the city and the ring road, the building is surrounded primarily by parkland. Local shops with converted loft apartments are located opposite, increasing the sense of a local community.

13.2 Brief

The project was awarded through competition. The key elements of the brief were to build forty-five one-bedroom flats of 50 m² each, with a total development-build cost of £50,000 per flat. The flats were to achieve a National Home Energy Rating of 9 (1). The NHER is based on the total energy running costs per m² per year, including energy use for heating, hot water, cooking, lights and appliances. A dwelling will receive a rating from 1 to 10. The development was also to be to the Joseph Rowntree's own Lifetime Home Standards (2) – a higher standard of accessibility than Part M of the Building Regulations (3).

13.3 Site-planning issues

A. Site plan

The site lies at the intersection of North Street with the inner ring road and is formed in a curve by the slip road joining the two. The building follows the curve of the slip road, thus completing the urban block formed when the motorway was constructed. To the south is a listed building, Centenary House, which has Victorian street elevations. The views from the site are from north-west to north-east across the slip road, and towards a small park, from which direction comes the principal sound impact of the motorway. Figure 13.1(a) shows the site plan.

All forty-five flats, representing 325 habitable rooms per hectare, are placed in a single block of between three and five storeys in height, following the considerable slope on the site from west to east. The flats have their principal windows facing south into the landscaped courtyard. Figures 13.1(b) and 13.1(c) show, respectively, the building elevation and street perspective.

Figure 13.1
Site plan and elevations
a. Site plan
b. Elevation at position P1
c. Street perspective at position P2

a

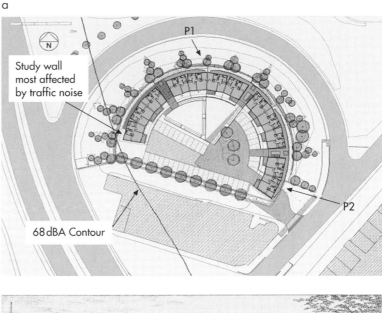

b

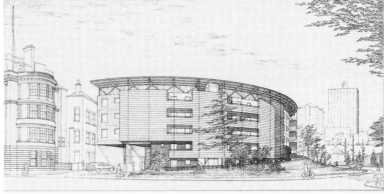

c

B. Noise

The site is on the ring road in Leeds and within the $L_{A10, 18h}$ road-traffic noise contour of 68 dBA (that is, the A-weighted noise level of 68 dBA is exceeded for 10 per cent of any 18-hour period). The external fabric of the flats within this noise contour forms the acoustic barrier to noise from the ring road. To provide an acceptable acoustic environment within the flats, noise levels of NR25 (roughly equivalent to 31 dBA – see Glossary) in the bedrooms and NR30 (roughly equivalent to 36 dBA) are the recommended design levels to avoid noise nuisance within the flats (4, 5). The brief called for the noise level within the flats to be less than 50 dBA with the windows shut. It was felt that this noise level was quite high and that it should be possible to improve on it. Guidance given by the World Health Organisation (6) states that "Where noise is continuous, the equivalent sound pressure level should not exceed 30 dBA indoors, if negative effects on sleep are to be avoided". In the end a design SRI (see Glossary) of 40 dBA was adopted for the external walls. This allowed a margin of 2 dBA to achieve the World Health Organisation's guidance figure of 30 dBA.

For the external wall to the study the sound reduction is achieved primarily with the mass of the wall itself. The SRI of the imperforate wall build-up is 45 dBA. With the introduction of a double-glazed window of SRI 33 dBA the overall SRI is reduced to 40 dBA. Triple-glazed windows were adopted, with a SRI of 40 dB, resulting in an SRI for the wall of 44 dBA. If a trickle vent were then introduced with an open area of 8,000 mm² (in accordance with British Building Regulations recommendations for a habitable room), the SRI would be reduced to 29 dBA, which would not have achieved the more stringent levels adopted.

Figure 13.2 highlights the weakness of penetrations in the acoustic performance of the wall. To avoid the need for trickle vents, fans were used to supply the fresh air and extract the stale air from the flats, thus reducing the area of opening in the wall. Acoustically lined ductwork further reduces external noise ingress.

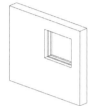
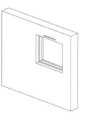

Figure 13.2
Sound performance of study wall options

Imperforate wall
SRI 45 dBA

Double-glazed window
SRI 40 dBA
Triple-glazed window
SRI 44 dBA

Triple-glazed window and
8,000 mm² trickle vent
SRI 29 dBA

C. Security

The courtyard has a single entry point from the street controlled by entry phone from each flat. From the courtyard, entry-phone-controlled gates lead to each of the three staircases and lifts. Accessible windows have key-operated locks and all windows and doors have multi-point locking. The site is secured with railings and galvanised mesh screens to the ground-floor access paths. The scheme achieved Secured by Design certification (7).

D. Parking

One space per unit was a requirement of the client's brief. Although Leeds planners would have preferred fewer spaces, the client's brief met the then-current Leeds Unitary Development Plan (8) for parking. In practice, less than 75 per cent of spaces have been taken up by residents, perhaps because of the additional charge levied on tenants requiring a parking space. This was a client-led policy to discourage car use on this inner-city site.

E. Sustainable urban drainage

Although a surface-water sewer was available to deal with rainwater it was decided to retain it on site, thus minimising the building impact on the sewerage system and avoiding problems of flooding elsewhere within the sewerage catchment. The subsurface geology consisted of between 0.8 m and 1.8 m of made ground, beneath which a strata of completely to highly weathered sandstone continued to underlying mudstone at a depth of approximately 20 m. Groundwater was not encountered during site investigation nor during a soakaway test; during the test the water soaked away as fast as it was poured into the sample excavation (9). Drainage from the roof was taken directly to two soakaways within the car-park and drainage from the car-park was taken to a third soakaway via a petrol interceptor. The paving to the car-park was predominantly $400 \times 400 \times 65$ mm precast concrete pavers, with butt joints, laid on sand, allowing surface-water to soak away locally and minimising surface water run-off.

13.4   Energy issues

A. Thermal fabric

A highly insulated external fabric was adopted with the following U-values:

Walls	$0.24 \text{W/m}^2\text{K}$
Door	$0.50 \text{W/m}^2\text{K}$
Roof	$0.20 \text{W/m}^2\text{K}$
Ground Floor	$0.20 \text{W/m}^2\text{K}$
Windows	$1.40 \text{W/m}^2\text{K}$

B. Heat loss

The fabric heat loss for a typical flat is 485 W (based on 20°C inside and −1°C outside). Internal heat gains from solar gain, lighting, appliances, cooking and metabolic rate of the occupants amount to approximately 450 W. The high levels of thermal insulation mean that the fabric heat losses will be substantially offset by internal heat gains. If the flats were naturally ventilated, via trickle vents, an air-change rate of 0.75 air changes per hour (ac/h) is suggested in the British Building Regulations Standard Assessment Procedure (10). This would result in a total flat heat loss of 1145 W. Allowing for the heat gain this represents a heat demand of 695 W.

C. Energy performance

In order to overcome the problem of noise intrusion into the flats, mechanical ventilation was adopted. This played a key role in the choice of heating system. The fresh-air supply can be heated to offset the heat losses from the flat. By running the fresh and exhaust air past each other in a sealed heat exchanger, half the outgoing heat can be reclaimed. Reference 5 recommends a minimum air-change rate of 0.5 ac/h for a mechanically ventilated dwelling. To introduce warm air at a temperature that is not too hot a ventilation rate of 1.0 ac/h was adopted. This ventilation rate (with heat recovery) plus an infiltration rate of only 0.1 ac/h at an external wind speed of 4 m/s, achieved by an airtight construction, plus the fabric heat loss less the internal gains, gives a total heating demand for the flat of 542 W. This is 22 per cent less than if the flats were naturally ventilated.

A kWh of grid electricity results in more CO_2 than a kWh of gas and so the effect of the fans must be considered. With heat reclaim and careful thinking on the control of the system it is possible to show that the CO_2 produced in heating the flats is less than that used by a conventional gas-fired radiator system with trickle vents providing ventilation. Using the calculation method in Reference 10 for space heating, the CO_2 produced by the CASPAR II system is 493 kg/y/flat and that by a naturally ventilated system with radiators is 593 kg/y/flat. This is a reduction of 17 per cent. The CO_2 produced by the heating of hot water amounts to 398 kg/y/flat.

D. System details

The heat-reclaim system is shown in Figure 13.3. The combination boiler provides hot water and heat to a water-to-air heat exchanger within the heat-reclaim unit. A room thermostat in the living room starts the boiler when there is a demand for heat. Because of the low heat load the boiler will cycle. (Combination boilers are specifically designed to cycle to meet intermittent hot-water demand.) The pipework on the heating system was oversized to provide sufficient thermal capacity in the water to limit cycling to within the manufacturer's recommendations.

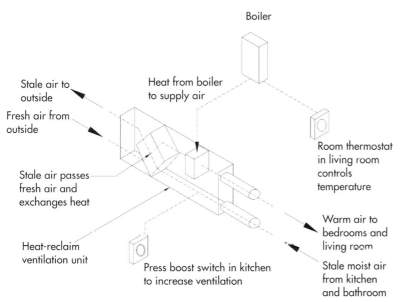

Figure 13.3
Schematic of heat-reclaim system

The design ventilation rate of one-half an air change per hour is achieved by cycling the fans at normal speed: one-half an hour on, then one-half an hour off. With a demand for heat the fans are run continuously at normal speed. When cooking or running a bath, a manual boost switch runs the fans at twice normal speed, automatically reverting to normal speed after 2 hours. With this control strategy the energy consumption of the fans is minimised. Figure 13.4 shows a layout plan of the heat-reclaim system within the flat. The airflows indicated are for the boost situation.

Traditional warm-air heating systems have a reputation for problems of low humidity resulting from the fact that the airflow rate is based on the heating requirement rather than ventilation. In the case of a leaky building, with high-infiltration heat loss, the introduction of large volumes of dry external air reduces the humidity in the space. In the CASPAR II system the low heat losses and airtight construction minimise the airflow rate so that the humidity within the flat is no worse than that of a naturally ventilated dwelling with radiators.

E. Air tightness

Fundamental to the success of the heat-reclaim system and the acoustic performance is minimal infiltration through gaps in the construction. The heating system was designed on the basis of an airtightness of 2 ac/h with a pressure difference of 50 Pa (equivalent to a wind speed of approximately 10 m/s on external walls) between inside and outside. The flats were tested under these conditions to ensure that this was achieved. Specifying a maximum air-change rate has little effect unless all potential air-leakage paths are identified and overcome through careful detailing. The pod units were wrapped in a waterproof membrane at the prefabrication factory

Figure 13.4
Layout of flat

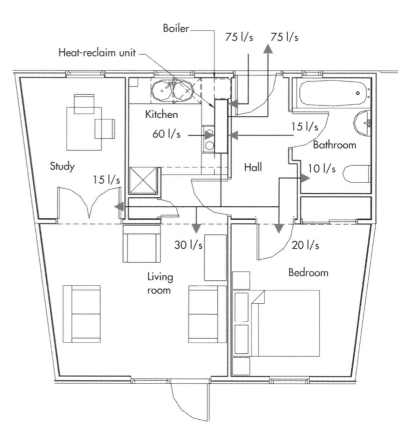

providing a barrier to infiltration. The triple-glazed windows are classified as Test Pressure Class 600 to BS 6375: Part 1 (11). This equates to $0.07\,m^3/h/m^2$ of window area at $50\,Pa$ pressure difference (with $6.4\,m^2$ of window, this is equivalent to $0.45\,m^3/h$ or 0.004 air changes per hour). Windows were detailed to minimise infiltration and letter-boxes were located at ground level to avoid penetrating the front door to the flat. Figure 13.5 shows a typical window-cill detail.

Figure 13.5
Window-cill detail

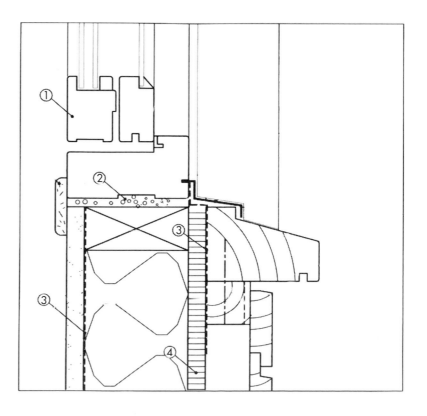

Key
1.   Treated-softwood triple-glazed window
2.   Expanding-foam filler
3.   Air- and vapour-check membrane
4.   15 mm bitumen impregnated fibreboard.

13.5 Construction issues

A. Design

To minimise the space taken up by the heat-reclaim system a bespoke unit was designed to fit above the kitchen wall units (see Figure 13.6). The inlet terminals were also designed to ensure that the air could be introduced with sufficient speed to ensure mixing of the air within the room. Without sufficient air movement the warm air would tend to rise to the ceiling and stagnate with cooler air at floor level, potentially causing uncomfortable temperature gradients. The terminal also allowed manual ventilation control by the tenant.

Figure 13.6
Heat-reclaim unit

B. Sustainability

The building's structural frame is comprised of multiples of 38×90 timber studs, clad with plasterboard internally and treated-softwood ship-lap cladding externally. Brickwork was used at ground level near Centenary House for contextual reasons. Wherever possible, water-soluble finishes were specified to minimise the release of VOCs (volatile organic compounds). The timber cladding is protected from rainfall by the stainless-steel-clad roof, which overhangs some 1.8 m along the front and rear elevations of the building. This enabled the entire timber external skin – except the two gable ends – to be designed not to receive rainfall, except under extreme conditions of driven rain.

C. Balconies

Each flat has a generous south-facing balcony to provide residents with an essential connection with the outside. The balconies are suspended on steel hangers from the cantilevered roof beams.

D. Wheelchair access throughout

Every flat is accessible to people in wheelchairs or parents with buggies via the three electric lifts (these lifts have been specifically selected so that motor rooms are not required). Three lifts were needed to minimise the length of the access galleries, thus avoiding unacceptable loss of privacy whilst maintaining the passive supervision of the galleries. The division of the flats into clusters grouped around staircases creates a collegiate sense of identity and security.

E. Radical approach to incombustibility (fire protection)

To minimise cost and for aesthetic reasons the steel frame supporting the access galleries was not treated with intumescent paint. Instead, fire-engineering analysis was undertaken by Warrington Fire Research to demonstrate that a fire in a single compartment (flat) would only sufficiently weaken one structural column, and to overcome this the structure was designed to span (under reduced loading) between three adjacent columns. The timber structure is protected internally by Fireline board.

F. Factory production (half the usual contract period)

The factory prefabrication system employed is known as "semi-volumetric". Fully finished "cores" comprising kitchen, bathroom, hallway and second bedroom, complete with fixtures and finishes, all doors, HVAC and other services were constructed off site in a factory in Bedfordshire. These were delivered by lorry to the site in Leeds, where they were craned in to place. The living rooms and bedrooms were constructed from traditional "flat-pack" prefabricated timber panels. The external cladding was designed to have been factory applied but was in fact carried out on site for cost reasons. A 36-week contract period was achieved – a significant reduction over traditional methods. The savings generated were ploughed back into the innovative features described above.

Figure 13.7
Layout of flat

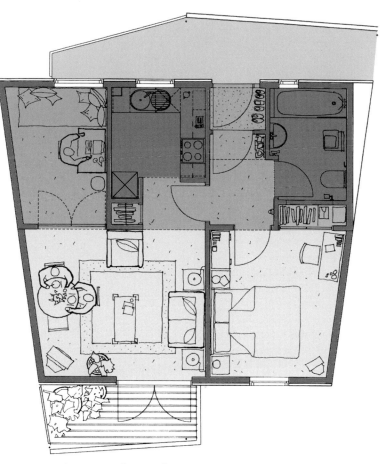

One/two-person flat (50 m²)
Flexible layout: one/two bedrooms

Living room/hall	19.4 m²
Main bedroom	10.8 m²
Kitchen	7.5 m²
Bathroom	4.9 m²
Bed 2/dining/study	7.4 m²

G. Lifetime homes

The Joseph Rowntree Foundation's Lifetime Homes standards were developed to encourage housing developers to prioritise access considerations, with a view to providing housing accessible to a wide range of users, from young people, people with families, to older or infirm people. It should be possible for a resident of a lifetime home to remain in their home regardless of the onset of old age, illness or accident. At CASPAR II, the following lifetime homes features were included: level access to all flats and thresholds; 1,000 mm external doorsteps; 900 mm internal doorsteps; wheelchair-turning circles in all rooms; walls and ceilings pre-strengthened for grab rails and hoists in certain locations; bathroom layout allowing for frontal and side transfer to WC; switches, sockets and controls located at easily accessible heights; lift access to all floors; and external works designed with ramps to a maximum of 1:15 falls.

13.6   Further development of the heat-reclaim ventilation system

The same heat-reclaim ventilation system has been installed at the Wilton Road development in Reading by the Ealing Family Housing Association. Half-hourly recordings of internal and external temperature and humidity, along with electricity, gas and water consumption, are being made for a typical flat. Surveying of residents via questionnaires is also being carried out. Preliminary findings from the questionnaires show that people have a desire to open windows even if their ventilation requirements are being met. The option of a manual switch to turn the heating system off, with a humidity-controlled override, is to be considered. A more refined method of controlling the fan speeds is also to be developed to reduce their power consumption and therefore the CO_2 they produce.

13.7   Conclusion

The success of the CASPAR II development has demonstrated that many new and innovative approaches can be adopted to produce a building that is sustainable in its energy use and can be procured within programme and budget. The final total development build cost was £51,100 per flat. The small overbudget cost of £1,100 per flat was partly due to unforeseen ground conditions. The heating and ventilation installation was supplied and installed for the same cost as an alternative radiator heating system with bathroom and kitchen extraction fans. Attention to acoustic detailing and the appropriate choice of ventilation system have enabled housing development within a noisy inner urban environment.

Project principals

Client	The Joseph Rowntree Foundation
Architect	Levitt Bernstein Associates
Services Engineer	Max Fordham LLP
Contractor	Kajima UK Engineering Ltd
Structural Engineer	Alan Conisbee and Associates
Quantity Surveyor	Robert Lombardelli Partnership
Electrical and Mechanical Sub-contractor	Elequip Ltd
Heat Reclaim Unit Manufacturer	Macklow Industrial Ltd
Timber Prefabrication Contractor	Volumetric Limited

REFERENCES AND NOTES

1.   Anon. (1993), NHER: A Guide for Housing Associations. UK.

2.   Carroll, C., Cowans, J. and Darton, D. (1999), Meeting Part M and designing Lifetime Homes, published by the Joseph Rowntree Foundation. York Publishing Services, York.

3.   Anon. (2000), The Building Regulations, Approved Document M: Access and Facilities for Disabled People. UK.

4.   Anon. (1968), Measurement and Control of Noise. EEUA Handbook No. 25, UK.

5.   Anon. (1994), Continuous Mechanical Ventilation in Dwellings – design, installation and operation. Building Research Establishment Digest 398, Garston.

6.   Anon. (1999), Guidelines for Community Noise. World Health Organisation.

7. Secured by Design (SBD) is the corporate title for a family of UK national police projects involving the design for new homes, refurbished homes, commercial premises, car-parks and other police crime-prevention projects.

8. Anon. (2001), Unitary Development Plan. Leeds City Council, Leeds.

9. Anon. (1991). Soakaway Design, BRE Digest 365. Building Research Station, Garston.

10. Anon. (1998), The Government's Standard Assessment Procedure for Energy Rating of Dwellings. Department of Environment, Transport and the Regions, UK.

11. Anon. (1989), BS 6375: Part 1. Performance of windows. Classification for weathertightness (including guidance on selection and specification). British Standards Association, UK.

FURTHER READING

Frairs, M. (2001), Is prefab just a fad? Building (9), pp. 24–26.

Powell, K. (2000), Building Study. The Architects' Journal, 3 (10), pp. 28–37.

WEBSITES

Joseph Rowntree Foundation: www.jrf.org.uk

Leeds City Council: www.leeds.gov.uk

National Home Energy Rating: www.nher.co.uk

Secured by Design: www.securedbydesign.com

World Health Organisation: www.who.int

14

Coin Street Housing: the architecture of engagement

Graham Haworth

14.1 Introduction

The architect's role in the development of the sustainable city is often ambiguous. Traditional concerns of architectural style, taste and composition preoccupy most good architects. The resulting designs, whilst visually provocative, are often problematic, unable to survive the closer scrutiny of wider socio-economic, environmental and cultural imperatives.

Equally, a simple commitment to social well-being through technology, as prescribed by orthodox modernism, avoids those situations where architecture has to engage with less-precise areas of reality. Architecture, as taught and published, often deals only with abstract certainties and this "retreat" is becoming increasingly unsustainable. The potential benefit of good architecture is denied to the wider population, and the product that we all get excited about in the media probably influences less than 5 per cent of all global construction, leaving unanswered the question of who is responsible for the remaining 95 per cent.

To be truly sustainable, architecture should be able to meet more objectives, satisfy more needs, be more appropriate and achieve more relevance. If it is to contribute to meaningful improvements in the urban environment it must engage with the "junk" of everyday reality and establish key values – values that are often generated as much by economics, social goals and politics as they are by design (see Figures 14.1(a), 14.1(b) and 14.1(c)).

Frank Gehry identified this in an early article entitled "Getting Tough With Economics" (1), in which he observed a discordance between designers' aspirations and the reality of the context they were working in:

> if you walk out on the street there are lots of cars, lots of dumb walls – this other thing called Design is a sort of forced attitude, it demands to be made of fancy not reality – the values are wrong.

As an architecture studio that is engaged in producing built work, our methods naturally focus on ways of contributing to a more sustainable urban environment. Embracing technology is essential, but we are sceptical of the reductive fetishisation of technology. Similarly, to predicate sustainable working practice exclusively on quantifiable indicators is oversimplistic, for the true measures of sustainability are also qualitative and cultural. We are interested in the pursuit of a humanistic culture; one that encompasses not only architecture, visual arts, literature and philosophy, but also the commonplace activities of daily life.

Establishing key values can take many different routes, but what increasingly interests us are those ideas away from mainstream thinking that exist on the periphery – "in the margins" – ideas that can bypass the "style, taste and technology debate". We are interested in extending the context of our work to engage with a range of cultural values, recognising the potential for architects to engage with the various agencies at work in creating the urban environment to achieve more sustainable results.

Figure 14.1(a)
Mission Street, San Francisco – the "junk" of everyday reality

Figure 14.1(b)
Manhattan – even "designed" cities cannot deliver the abstract certainties desired by design professionals

Figure 14.1(c)
Manhattan – "the seventh investigation", 1969, by billboard artist Joseph Kosuth explores the non-art context of the public realm

Figure 14.2
The paraSite shelter for the homeless by Michael Rakowitz

Figure 14.3
Gemini GEL by Frank Gehry – adaptable, low-cost, low-tech cut and paste

Figure 14.4(a)
Jae Cha's low-cost community projects in Honduras (top) and Bolivia (bottom)

Figure 14.4(b)
"Maximum space for a minimum cost" is the sustainability maxim upheld by this Will Bruder makeover in Arizona – US$10 per m² "rewrap"

Consider shelters for the homeless in New York, for example, designed by Michael Rakowitz, a 27-year-old "artist, instructor, designer and activist" (see Figure 14.2). Rakowitz made his first structures from garbage bags and contact cement as a graduate student at MIT and has recently developed an ongoing customised shelter project he calls "paraSite". Temporary, inflatable structures provide life-sustaining warmth and shelter by taking in the wasted exhaust heat of air ducts and steam vents. They cost US$5 to make. As Rakowitz points out, his work should not really exist and the shelters should disappear along with the problem. "In this case, the real designers are the policy makers and the ones who are capable of organising bureaucracy" (2).

An early Frank Gehry project for Gemini GEL, a company that produces fine-art graphics, takes the idea of the simple stucco box of neighbouring light industrial buildings, with their flat roofs, packaged air-conditioning and chain-link parking lots, and hacks it around, creating a low-cost, dynamic urban environment, responsive to future replenishment, addition and change. Gemini was originally two small buildings, which Gehry remodelled and clad with a new façade in grey stucco. A 5,000 square foot two-storey gallery and workshop were added subsequently and the car-park was later developed as a restaurant (see Figure 14.3).

Jae Cha's low-cost work in Bolivia and Honduras uses the simplest of materials – timber and polycarbonate sheeting – in beautifully poetic ways. The results are extraordinarily powerful, demonstrating an exemplary understanding between architect and community (see Figure 14.4(a)). In Arizona, Will Bruder's radical remodelling of a typical suburban house regenerates an existing structure through an economic lightweight "rewrap" and is a perfect model for inventive sustainable urban regeneration (see Figure 14.4(b)).

What is inspiring in this kind of work is that while it concerns both aesthetics and technology, the key innovation, the starting point for the idea, is often economic, social or political. We have recently undertaken several projects in London that deal with similar issues and have helped us focus more laterally on ideas about sustainability and what it really means. This has clarified our thinking on how the process of making architecture can enhance sustainability, particularly in an urban environment, and is essential in providing both the catalyst for change and the glue that binds together the results of that change.

14.2   Iroko Housing Co-op for Coin Street Community Builders

One of the projects, for Iroko Housing Co-op is explained in depth here as a way of illustrating the complex issues at play in sustainable urban regeneration and how certain criteria that architects often consider to be peripheral can positively influence the design.

Iroko represents the latest stage in the development of a group of sites, totalling approximately 35 hectares, on the South Bank of the Thames. The client, Coin Street Community Builders (CSCB), is a non-profit-making company that aims to provide affordable homes in the centre of the city, and over the past decade CSCB has transformed this part of London.

Located immediately behind the National Theatre, one block back from the River Thames, Coin Street has become a model for community-led urban regeneration. There will eventually be some 350 homes at Coin Street, owned and managed by six separate housing co-ops, together with arts and community facilities, shops and workshops, cafés and restaurants, a park and a riverside walkway (see Figure 14.5).

Coin Street are genuine "community builders" – constructing the framework for a sustainable community on the South Bank. Importantly, they are not just looking at the physical design of housing in isolation, but at forming the socio-economic structures needed to allow residential accommodation to integrate with other uses. These range from local initiatives such as festivals, education, sports and community activities (see Figure 14.6), but also involve forging links with prominent cultural institutions such as the National Theatre, Hayward Gallery and Royal Festival Hall, together with new major employers such as London Weekend Television and Sainsburys.

The project, won through a limited-entry competition in 1997, provides a total of fifty-nine dwellings and includes thirty-two family houses, each of which accommodate up to six people; the balance of accommodation is made up of a mix of flats and maisonettes. All dwellings are for rent and will be managed by a housing co-op formed by the residents. The total site area is 1.2 hectares, with 0.8 hectares for housing and the remaining 30 per cent of the site area reserved for a community facilities building, "The Hothouse" which forms the major mixed-use element of the scheme.

Formerly occupied by several large warehouses about six storeys in height (see Figure 14.7), the site was cleared to basement level over 10 years ago and used in the intervening period as a car-park. The brief called for the public parking to be retained and, as the basement walls remained, it was an obvious decision to place the new car-park below ground and develop the housing from ground-floor level upwards.

The cost of the housing substructure could then be cross-subsidised by the car-park, effectively providing the housing with foundations and ground floor for no cost. Detailed design of the housing and car-park layout was carefully co-ordinated to avoid expensive transfer structures wherever possible and many of the existing concrete bases from the old industrial buildings were used for new foundations.

This site was one of the few in the area that could accommodate large family homes in the form of individual houses, with adequate levels of outdoor-amenity space. Initial designs illustrated that between fifty and seventy dwellings, containing thirty to forty houses, might be achievable. Conscious that social housing tends to be fully let and that over-high child densities on estates have been shown to create problems, Coin Street emphasised the desire for a spread of dwelling sizes, confirming that its prime concern was quality of accommodation and environment rather than maximum density.

The resulting density was therefore design-led, although it does coincide with new government guidelines set out in PPG3 that propose higher housing densities on urban brownfield sites to protect London's green belt (3). The density of the project is 291 habitable rooms (334 habitable rooms per hectare), an increase of 59 per cent on Lambeth's planning guideline of 183 habitable rooms for the site (or 210 habitable rooms per hectare). The development is not yet at full capacity. There are at present 260 occupants (298 per hectare), of which 140 are children under the age of 16. If fully occupied, this would rise to 360 occupants (413 per hectare).

The site is interesting in a number of ways as it straddles two very different worlds. While Waterloo International Station and the nationally important cultural area of the South Bank centre are immediately to the north and the west, the traditional residential neighbourhood of Waterloo and The Cut defines the southern and eastern edges (see Figure 14.8).

The river is not present but is accessible just off Upper Ground via a number of pedestrian links. All four boundary conditions on the roughly square site

Figure 14.5
The Coin Street area seen from across the Thames with the South Bank centre and Waterloo Station in the background

Figure 14.6
Bangra workshop at one of Coin Street's summer community festivals

Figure 14.7
The site was formerly used for warehouses and factories, supported at the time by the Thames as a "working" river

Figure 14.8
Looking west from the Coin Street area, with the recently completed first phase of Site B in the centre of the picture

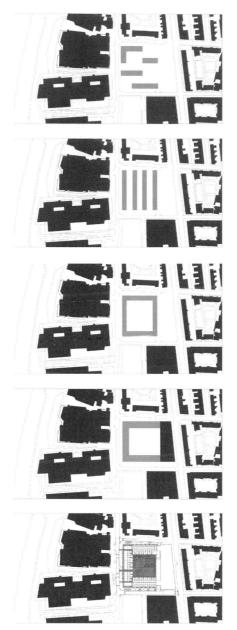

Figure 14.9
Various configurations for housing layouts at Coin Street

Figure 14.10
Stanley Gardens in Notting Hill

are very different in scale and feel. The most dominant adjacent buildings are the twenty-storey London Weekend Television tower and Cornwall House, a seven-storey former warehouse building recently converted to an educational facility for King's College, London University.

City-centre housing still proves to be something of a paradox. A sufficient scale of architecture is required so that the buildings can fit comfortably into a metropolitan context, but the housing must also provide a setting for small-scale domestic activity; therefore, finding an appropriate scale of response to the site environment was the main starting point for the design. We were also conscious that the status of the inhabitants of European cities had undergone a sea-change in the previous decade. Urban dwellers are no longer marginalised, deprived or socially excluded; they are confident, creative and self-assured. But in parallel with the rise in confidence of the individual, there has been a demise in the collective confidence of the community, which we felt could be addressed in the design.

The balance between the monumental and the domestic was examined as a way of creating a specific sense of place and community, with a unique identity, through an architecture that had sufficient scale and presence to respond to the exposed urban context but at the same time could create high levels of privacy and communal amenity for the residents. We believed that on this site a strong typology was needed that could be easily read and understood by both the public and the residents — a form that established very clear signals of public and private.

We explored various configurations for housing layouts, including traditional street patterns and individual housing blocks, most of which were found to be unable to provide the right level of defensible space. The final form adopted was the hollow-square model. This helped to redefine the urban block as a whole and provided the right level of external enclosure, leaving the internal space open as a central communal amenity. The dwellings are arranged around an open courtyard, which allows communal space to be maximised in the form of a large, landscaped garden with play-areas, bounded on three sides by residential accommodation and on the fourth side by the community-facilities building, which is to be constructed as the next phase of the development (see Figure 14.9).

The hollow square is a form used most frequently for multi-storey apartment buildings where the habitable rooms of the dwellings face outwards to the street and where access takes place on the internal face. The use of a hollow square for single family dwellings is less common but some of the best examples are in the garden squares of Notting Hill, where private gardens open on to a communal garden that is shared by occupants from opposite ends of the site. This model has proved very successful, the central space becoming a focus for shared ownership and interaction (see Figure 14.10). We were also interested in the calming effects of landscape and planting in the inner city (see Figure 14.11).

Figure 14.11
The communal garden at Bonnington Square in Vauxhall is an escape from the busy roads surrounding the buildings

The scale of the design proposals is a response to the inner-city environment, which demands a particular massing and density to maintain the metropolitan buzz. The fact that the project is constructed over a concealed basement car-park for 265 cars adds to the urban concentration. The massing of the housing reflects the urban setting and acknowledges the context of the streets it addresses. The houses in Coin Street and Cornwall Road are three storeys high with an "attic" rooftop room at fourth-floor level. This is set back from the street elevation and opens on to a rooftop terrace on the courtyard side (see Figures 14.12 and 14.13).

The houses in Upper Ground are three storeys with two-storey maisonettes above, accessed via a common external balcony, again overlooking the courtyard. The increased height here is a response to the busy urban thoroughfare and the mass of the London Weekend Television Centre on the opposite side of the road (see Figure 14.14).

A simple series of principles was established to plan the layouts of the dwellings. All the houses have street-level entrances and private gardens opening on to the communal courtyard. All the flats and maisonettes have large balconies and each of the bedrooms overlooking the courtyard has a balcony (see Figure 14.15). The façades of the buildings acknowledge their dual aspect, addressing the public streetscape on the outside and the private landscaped garden space in the centre of the block. The street elevations are expressed as a brick screen with deep window reveals and the elevation steps back at the upper-storey levels, clad in vertical pre-weathered zinc (see Figure 14.16).

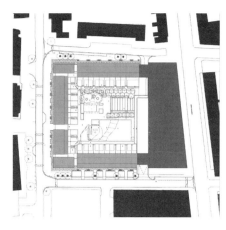

Figure 14.12
Site-strategy diagram for the whole development

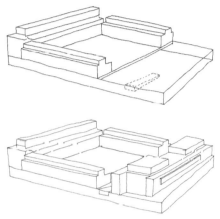

Figure 14.13
Massing diagrams

Figure 14.14
View into the communal courtyard garden from the "Hothouse" site

Figure 14.15
Corner configuration of housing units at upper-terrace level

a

b

Figure 14.16
Inner courtyard elevation (a) and streetside elevation (b)

Figure 14.17
Landscaped courtyard entrance

Figure 14.18
View of the surrounding urban context
from upper-terrace level

Figure 14.19
The courtyard is separated into distinct
usage areas, defined as hard or soft
landscape

The party wall between each dwelling is expressed on elevation so that each dwelling is easily identifiable from the outside. The individual unit is recognisable but clearly forms part of a larger whole. The rigorous design of the street façades is intended to provide a strong and ordered backdrop against which the inevitable personalisation of net curtains and potted plants can take place.

The houses have a "privacy layer" at street level in the form of a raised pavement area separated from the main public thoroughfare by stainless-steel and timber bin enclosures and architectural metalwork. Each house also has a porch area immediately off the street, enclosed with a metal gate and housing a large cupboard area for the storage of buggies and on line shopping deliveries.

The palette of materials on the courtyard side has been carefully selected to weather and mature with the landscaping. The infill cladding is timber – *Vitex cofassus* and *Lauro pretto*, both naturally durable hardwoods from managed sustainable sources. They require no preservative treatment or applied finishes, thus reducing maintenance costs. The same material is used for horizontal timber sunshades and balcony decking. On the garden elevations the concrete frame is expressed with the edges of slabs and the faces of columns on party-wall lines exposed. The primary cantilever members of the steel balconies are grouted into sleeves cast into the slabs (see Figures 14.17 and 14.18).

The landscaped courtyard has been designed to cater for a range of activities and age groups. Pathways, large planting beds and profiled concrete walls divide the space into four main areas: a large sloping lawn area with seating terrace, focused on what will grow into a specimen tulip tree; a toddlers' play-area with play equipment; a sunken ball-game area with stepped seating platform; and an area in which a single-storey meeting building for the co-op will be located (see Figures 14.19 and 14.20).

The proposals embody many of the principles of sustainability relating to spatial planning and solar access – each dwelling has roof-mounted solar panels for the production of domestic hot water – and the specifications for insulation levels, ventilation systems and materials have been chosen to have the minimum environmental impact.

Each of the dwellings is fitted with mechanical background ventilation, which offers the residents maximum comfort control. Ventilation rates can be modified and heat recovered from the exhaust air to preheat the supply side of the system. Condensing boilers are used, which have heat efficiencies of up to 90 per cent, reducing heating bills by about 30 per cent.

The integration of other mixed uses with the housing is an important part of Coin Street's sustainable-development strategy. On this site the community facilities building will be used to close off the fourth side of the square and provide a visual and acoustic screen to Stamford Street, a busy traffic artery running between Waterloo Bridge and Blackfriar's Bridge.

There are very few indoor facilities that specifically serve the needs of Coin Street's growing residential community. In addition, employment in the area's traditional industries – such as the docks and the printing industry – has been in decline and is now being replaced by opportunities in the new growth industries of leisure, arts and media. However, many inner-London residents fail to secure employment in these growing industries because of inadequate training.

The Coin Street "Hothouse" has been developed to address both of these needs by creating partnerships with central London employers and by providing community and training facilities on the southern part of the site adjacent to the housing (see Figure 14.21).

Plans have been developed for a building containing meeting, bar and crèche facilities specifically for local residents, a cybercafe and IT resource centre, spaces for meeting, training, advice and rehearsal, and small offices for local groups and arts organisations, and also retail, catering and exhibition areas.

14.3 Background to Coin Street Community Builders

The origins of CSCB illustrate that true community involvement in both the planning process and urban regeneration initiatives can lead to more sustainable urban design. CSCB came out of a campaign that ran during the 1970s and 1980s and two year-long public enquiries overseen by the then Secretary of State, Michael Heseltine.

The Coin Street area was in a state of decline. Big employers such as Boots, WH Smith and Her Majesty's Stationery Office had moved out and their premises lay empty. The city as a whole was in a state of flux with high unemployment in traditional industries and the decline of the Greater London Council.

It is easy to forget the mood of that time. Riots in Brixton and Hackney and the voice of disenfranchised youth were perfectly captured by punk music, which saw London burning – a wasteland of tower blocks, heroin, police and thieves. Although the original residents found the area threatening and menacing, there was deep scepticism over the notion that displacing the local community with a commercial development would be a sustainable long-term option.

Instead of providing a clear statement on the planning needs of the area, the policy had been one of waiting to see what the market came up with. Such vagueness seemed to be the antithesis of good planning. Local residents, alarmed at the closure of local schools and shops, the loss of housing and the prospect of further office development, decided to get directly involved in the planning process (see Figure 14.22).

The sites became the subject of two separate planning applications, the more dramatic of which was designed by Richard Rogers for Greycoat Estates. Rogers' scheme was an ambitious single vision, a masterplan for a necklace of tall new buildings along a covered galleria, one-third of a mile long, stretching from Waterloo Bridge to the OXO Tower.

As a formal piece of urbanism it had great design integrity but relied on over 100,000 m² of commercial space to be viable. It also proposed demolition of the OXO Tower Wharf and the creation of a new pedestrian bridge link over the Thames. The scheme was aimed at a new city-wide public as an extension of the South Bank public amenity, but many felt that it neglected local issues.

An alternative scheme put forward by the Association of Waterloo Groups, later to evolve into CSCB, avoided such a grand statement and concentrated instead on a range of smaller ideas, starting with the needs of the indigenous community, reinstating the original street pattern of urban blocks and retaining OXO Tower Wharf.

Notwithstanding the differences in architectural approach, the issues were really about scale and use mix; the choice between a city-focused commercial scheme with an element of residential (Greycoat) or a locally focused residential scheme with an element of commercial (Coin Street). There followed one of the country's longest planning battles, which resulted in both schemes being granted planning permission.

Although the strength of opinion favoured Coin Street, local community groups would have been unable to afford to purchase the sites, which now

Figure 14.20
Hard-terrace areas within the landscaped courtyard

Figure 14.21
Street-side façades of the "Hothouse" community-facilities building

Figure 14.22
Local action-group protests during the public enquiry

had the benefit of planning permission for office use, without the direct intervention and support of the local authority, which, after much lobbying, imposed covenants on the site, restricting the purchaser to building fair rented housing and light industrial/shopping space in accordance with the community planning permission. This reduced the value of the site to a fraction of its true commercial market value, making it an unattractive proposition for commercial developers.

Every planning authority has such powers to intervene directly but seldom uses them. This controversial change to the UDP and the subsequent land sale by the disbanding GLC ended in the development of what was possibly "the capital's most valuable piece of real estate" being undertaken by a local community group whose scheme flew in the face of market forces.

At the time Coin Street was highly controversial but before long most people were beginning to wonder what all the fuss was about as the community scheme proved to be so right for the area. The sustainability lessons from this, which can apply to other city-centre sites, are quite simple.

It is possible to achieve social goals through new development and for a community to harness the power and support of the elected local authority to implement positive change. Clearly the involvement of the local groups in planning and development early in the process resulted in a more complex range of options being considered. The involvement of the community provides an invaluable source of local knowledge, history and cultural memory, the seed-bed from which more sustainable urban communities can grow (see Figure 14.23).

Figure 14.23
Ernie – a local resident and CSCB committee member, recently awarded an MBE for services to the community

The evolution of Coin Street also demonstrates that it is possible to avoid rigid, single-vision masterplanning and replace it with a more holistic approach to urban planning that, rather than forcing through one big idea, advocates the synthesis of lots of small ideas into a whole. Masterplanning can be diverse and provide a menu of projects that can be implemented by different architects as and when money and land become available. It is still possible to start out with a wide vision but enable it to be broken down into more manageable phases. This is more organic – it can adapt to change as communities evolve and develop in different ways than might originally have been envisaged.

Coin Street also illustrates the need for defined management and organisational structures that enable the initiative to develop and survive. The Coin Street area consists of seven different sites, all of which ultimately are to be developed predominantly for social-housing use, although several of the sites were zoned to provide for mixed use in addition to residential. OXO Tower Wharf, for example, has ground, first and top floors given over to commercial use, with seventy-eight residential apartments in between.

The main board of CSCB sets the overall development guidelines for the area and forges educational and employment links with local institutions and businesses. A management group, South Bank Management Services, runs the infrastructure, rents out the commercial space and organises festivals and cultural events. It even has its own team of gardeners to maintain the parks, riverside walks and other public areas. There is a clear separation in management between the infrastructure and the housing.

Each element of the residential is run by a specially formed housing co-op and there are now four of these in existence, all, optimistically for an urban setting, named after species of trees – Palm, Redwood, Mulberry and Iroko. A housing co-op is a group of people who jointly own their homes and control the way the housing is run. The properties belong to the co-op as a whole and members pay rent to the co-op.

The co-op is non-profit-making, so any surpluses go back into improving the housing. This keeps down management costs and rents, particularly if members do some of the day-to-day work themselves on a voluntary basis. The co-op system also provides on-site management so that the usual problem areas of high-density social housing, such as supervision of communal spaces, lifts and stairs, can be adequately controlled. CSCB provide a training programme that helps the residents to understand their responsibilities as co-op members so that they feel confident in carrying them out (see Figure 14.24).

Figure 14.24
The courtyard landscape acts as social focus for the residents

Unlike most social housing groups, CSCB have developed a team with a range of skills, enabling them to exploit the huge potential of the sites and their location to provide opportunities to cross-subsidise the housing. The OXO Tower contains two large up-market restaurants and a series of craft workshops, a coffee bar, a florist and gallery spaces. The 265-space car-park beneath Iroko has been leased to NCP and provides spaces for local office workers and visitors to the South Bank. There are also strategically placed advertising hoardings and illuminated billboards, which bring in much-needed revenue.

Some people find it difficult to understand why a not-for-profit company like CSCB should want to entertain these commercial activities, and lease a restaurant to Harvey Nichols that serves fairly expensive food to wealthy people, and provide spaces for commuter parking when they should be discouraging the use of cars in central London. The answer is purely pragmatic. They make money from the commercial parts in order to fund those other parts such as maintaining the parks and gardens, having festivals and cross-subsidising the housing. The commercial reality of the cross-subsidy is essential as building high-density inner-city housing is on average 100–150 per cent more expensive than building traditional two-storey housing.

Further phases of Coin Street are currently at the brief-development stage to form the basis of architectural design competitions, which will ultimately provide further housing, indoor sports facilities, an Olympic-size swimming pool and the "Hothouse".

What can we learn from Coin Street? Primarily, for us it is another example of "working in the margins" – challenging conventional wisdom on the role and practice of architecture. Several characteristics of CSCB's approach establish valuable standards for other sustainable urban communities:

- the origins and background of CSCB itself;
- positive involvement of the local community in the planning process;
- adoption of neighbourhood masterplanning;
- co-op-based structure of the organisation and community involvement in the design process;
- creative use of cross-subsidy and funding;
- commitment to high build specification and high initial capital investment;
- diversity of design;
- high density and mix of uses;
- low energy use and use of renewable energy sources.

All these characteristics provide measurable yardsticks for a sustainable community, and we envisage that in the future we might see a series of similar initiatives, which draw inspiration from Coin Street and its achievements in bringing about positive change. They will probably be different in scale, materiality and diversity, and there is likely to be more emphasis on improvisation, on the temporary and on the provisional in the city.

We are experiencing this shift already and are increasingly involved in an examination of found or "un-designed" space as a source of sustainable

Figure 14.25
The Almeida Theatre at King's Cross

working methods. One of our recent projects, a temporary theatre for The Almeida at King's Cross (see Figure 14.25), explores these possibilities, using found space as a starting point and enclosing the maximum amount of space for the minimum cost in the most economic material. The sedum landscaped roof was not even paid for but was rented from the supplier, who will take it back at the end of the project for use on another building (see Figure 5.16).

In addressing contemporary issues of sustainability, the poetic of the temporary and the improvisational becomes increasingly attractive, encouraging an architecture that borders on "non-design", is provisional, and is open to addition and change. An urban architecture is evolving that celebrates ambiguity and improvisation, underpinned by humanistic values – a type of "post-net" vernacular. The main challenge for the architect in contributing to a sustainable city will be how to balance the force of this inevitable change with the more qualitative values of cultural memory and history.

Project principals

Client	Coin Street Community Builders/Coin Street Secondary Housing Co-operative
Architects	Haworth Tompkins
Services Engineers	Atelier Ten
Structural Engineers	Price & Myers
Quantity Surveyor	Davis Langdon & Everest
Landscape Architects	Camlin Lonsdale
Project Managers	Haworth Tompkins
Main Contractors	The Mansell Group
M & E Contractors	Briggs & Forrester

REFERENCES

1. Soukler King, C. (1982), "Getting Tough with Economics", interview with Frank Gehry. Designers West, June.

2. Chen, A. (2001), "Tent City", interview with Michael Rakowitz. i-D Magazine, February.

3. PPG3 (2002), Planning Policy Guidance, note 3, Housing. Department for Transport, Local Government and the Regions, March. This document provides guidance on a range of issues relating to the provision of housing. It replaces the 1992 version of PPG3.

FURTHER READING

Anon. (2001), Home: a place to live, an oasis of tranquillity in the heart of London. The Housing Design Awards Catalogue.

Glancey, J. (2002), Heart of Gold. The Guardian, 8 April.

Gregory, R. (2002), Case for the intense. Building Design, 8 June.

Hinsley, H. (2002), Architecture Today, April.

Powell, K. (2001), Housing, Coin Street, SE1. New London Architecture, Merrell.

Spring, M. (2002), Coin Street Housing. Building, issue 10.

Worsley, G. (2001), Designer buildings for a new model community. Telegraph Property, 10 November.

Sustainable design in an urban context: three case studies

Alan Short

15.1 Determinism in sustainable design

Fifteen years ago it seemed as if passive solar design, as a conscious intent, forced building form into mono-pitched prisms oriented carefully on plan. Rigorously applied "heliotropic" site planning organised groups of prismatic objects parallel to each other and spaced apart to maximise solar exposure. Try, as one might, it was very difficult to induce such a strategy to make a recognisable urban environment. In the 15 years we have spent trying to design sustainable buildings, the majority of which have been in some kind of urban context, we have discovered, slowly, that with the application of ingenuity and imagination one can transform a building with almost any shape and orientation into an effective functioning passive solar object. This chapter tracks the development of our thinking, principally in plan form, from the wide-frontage, narrow-section buildings of then, to the deep and condensed plans we have been developing more recently. Figure 15.1 maps this slow development. Twelve years ago, whilst developing the plan of the New School of Engineering and Manufacture for Leicester Polytechnic, we speculated that the strategy developed to enable a building to interact more directly with its "natural" context in order to reduce its consequent carbon dioxide emission might simultaneously help to embed the building more meaningfully into its physical, urban context. The new school, later to become the newly incorporated De Montfort University's Queens Building, is a 10,280 m² laboratory building, a type for which the twentieth century prescribed a standard diagram of a deep-plan, cubic building, with three or four floors of double-loaded corridors, enclosing a larger, central volume, all sealed and mechanically ventilated and cooled, and largely artificially lit. As a prompt, upon appointment, we were shown drawings of such a building recently completed in Sweden by the then Head of Engineering. We were intrigued as to how the application of a particular environmental strategy could help to dissolve this rather large building and its unforthcoming standard plan type into its complicated context. As we completed the Queens Building the opportunity arose to consider the integration of a sustainable theatre into a yet denser urban fabric.

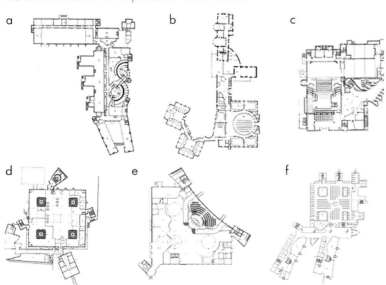

Figure 15.1
Comparative plans of the Queens Building (a), University of Manchester Conference Centre (b), Contact Theatre(c), Coventry University Library (d), Thames Valley University New Learning and Teaching Building (e), Judson College New Academic Centre (f)

15.2 The Contact Theatre, Manchester: the context in 1993

In 1993 the Contact Theatre Company was a full producing house operating on a shoestring with a company strength of sixty-five people in two buildings separated by 100 metres of windswept temporary university car-park. The simple rectangular auditorium building with an exceptional stage width of 17 metres was built in 1963. It had very thick load-bearing masonry walls but a lightweight roof of channel-reinforced woodwool slabs, felt-covered. Its roofline pitched gently up along the longitudinal axis of the auditorium and some sort of operating height was recovered economically over the stage without any distinct discontinuity in form. The side cheeks of the tilted mansard were faced in copper. A small foyer on two levels was contained within the rectangular envelope to the south of the main auditorium across the full width of the rectangular plan, and a band of tiny dressing rooms and a sub-stage completed the north end of the box. A scene dock projected to the east with a large dock door. The blank scene dock formed, in effect, the main approach elevation to the theatre (see Figure 15.2).

Figure 15.2
The Contact Theatre in 1993

15.3 The site

The building completed the quadrangle of the Victoria University of Manchester Humanities block. Immediately to the west was, and still is, a ten-storey slab block.

The other three sides of the courtyard are variously of two and four storeys. This is the slightly forgotten south-west quadrant of the university campus and, as at Leicester, a new and very incomplete grid of 1960s framed University buildings was set off across the quite differently oriented tighter pattern of nineteenth-century Victorian terrace housing. Demolition here, however, was even more enthusiastically carried out than was that across the west side of Leicester. Few of the houses remain, streets stop and start in an apparently random, but syncopated way and one can track them, disappearing and reappearing beneath faculty buildings – a postman's nightmare. In 1993 the company's perception of itself was that it was almost invisible from the Oxford Road, the bustling, brightly lit north–south artery of the university campus. They confidently expected the university to develop the acres of Tarmac poured around them into new university buildings, thereby completely squeezing the theatre out of the public realm. The company's headquarters was a decommissioned pre-war, single-storey school building in which it made all its scenery and props, administered and marketed itself, designed and planned new productions, rehearsed them, and even presented them in its classroom studio venue. Actors and crew trudged backwards and forwards across the car-park carrying scenery and/or fully dressed to make an imminent appearance on stage in the main auditorium building.

15.4 The company's aspirations

The company had become highly respected by the end of the 1990s and was known in some circles as "the Young Vic of the North", despite these almost unworkable conditions. Furthermore, this quadrant of the campus connected with the Moss Side and Hulme estates and so qualified at its margins as an exceptionally degraded inner-urban area on the European map of urban decay. The task was to gather the company together and put them in one building with a main auditorium, a studio theatre for their Young Person's Company and a rehearsal room that bore some resemblance to the shape and size of the main stage. The second task was to make a building that one could not possibly miss from the Oxford Road and, the hope was, from yet further afield. The third task was to relieve the theatre company of the potential considerable financial and operational burden of a large mechanical ventilation and cooling system (still *de rigueur* for new theatres and for major performance-venue refurbishments).

15.5 The company's green intent

Publicly funded theatre companies in Britain tend to have no spare cash, by definition, and certainly cannot employ full-time building managers. The original mechanical-ventilation scheme was soon found to be so noisy that it was literally never used. During performances temperatures rose in the original unventilated auditorium to almost intolerable levels, even during pre-Christmas pantomime performances when outside temperatures were near freezing. The auditorium was almost completely unventilated during performances and theatrical smoke would hang in the air without any motion at all, causing upset to subsequent scenes. The actors, members of the home team and the numerous visitors passing through became very exercised during our extensive winter of consultations about the difficulties of acting in an air-conditioned environment. We were somewhat taken aback by the strength of feeling. Our clients were many times more evangelical than we were about avoiding air-conditioning through judicious design. The actors explained that it was almost impossible to generate any kind of atmosphere in an air-conditioned auditorium and engage with the audience if the air, the medium for the exchange, was constantly refreshed and cooled. Marching about at the front of the stage in a heady on-shore cooling breeze was, to quote, "wholly destructive of the theatrical process" (1).

15.6 Naturally ventilating a theatre in an urban environment

Naturally ventilating and passively cooling a theatre is possibly the most complicated and demanding application of the idea, particularly in an inner-urban environment. The acoustician (2) discovered during his 24-hour

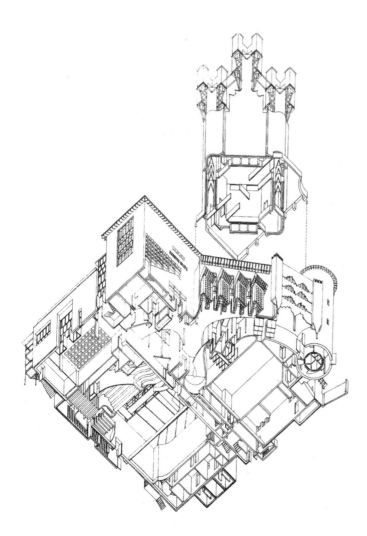

Figure 15.3
Choisy view up into the transformed theatre, showing the air-intake plane beneath the performance spaces

sound survey that the centre of Manchester is extraordinarily noisy. The sound level in dBA exceeded for 10 per cent of the time L_{A10} was measured as 61 dBA. Close to the Contact is a rock-music venue called the Academy, which is on the national touring circuit. On the other side of the Humanities quad was an audiology clinic on the ground floor and Humanities teaching happened all around the theatre, the most acute conjunction of uses being in the pre-Christmas pantomime matinée season. The revenue from the pantomime underwrote the company's annual budget. Acoustic issues rose to such prominence that they became very influential in determining the broad architectural solution. Not only was it highly desirable to keep external city noise out of all three venues but it was also a statutory requirement to keep the sound generated within all three venues, the workshop and the scene dock out of the city and the surrounding buildings, and, for obvious operational reasons, out of each other. The Contact was famous for the design of its stage sets, which tended to be wildly ambitious in both concept and form. It emerged that it was extremely important that the next programmed production could be constructed and assembled whilst the current production was running, without lengthy and costly programme-endangering interruptions.

15.7 The strategy

The site was very tight despite the Vice-Chancellor's generosity in allowing the serpentine street skimming the original building to be shifted east by about 8 metres. The key spatial relationship to achieve was the juxtaposition of the scene-assembly dock on axis with the stage of the main auditorium so that sets could be slid through on to the stage. The lateral dimension is very tight. The machine room containing electric saws and drills had to be placed immediately adjacent to the scene dock. The main auditorium was stripped out to its shell and its broken back-seating rake rebuilt in a continuous parabolic section. The 110-seat studio theatre was literally lifted off the ground two storeys and placed, floating on neoprene acoustic buffers, on top of the scene-assembly dock and the workshop (see Figure 15.3). The rehearsal room, similarly, left the ground and rose up over the foyers and administrative offices, almost colliding with the studio theatre two floors above the street.

Figure 15.4
Model testing in the wind tunnel at the University of Wales

15.8 Particular design issues

Basic air-flow patterns through the building

The more detailed design task was to superimpose closed loops of intake and extract in three dimensions to and from all three performance venues, and the foyer spaces between several levels. This had to be achieved whilst maintaining strict fire compartmentation and whilst defending the paths in and out against noise and sound coming in either direction, and against rain, wind and sleet. Wind-tunnel tests at the University of Wales (see Figure 15.4) revealed the wisdom of achieving an intake oriented to every cardinal direction to cope with the seasonal variability in wind direction.

Developing robustness to wind direction

South-westerlies in particular are prominent from March to September. The ten-storey Humanities tower lies directly to the south-west of the theatre. The wind-tunnel experiments revealed how crucial it is in naturally driven schemes to maintain pressure differentials between the inlets and outlets to support the internal airflows. The first schemes proved to be very vulnerable to a cessation of movement and even reversal of flow under certain wind directions. The models were modified on site to test a variety of approaches. The first model proposed an inline extract arrangement on top of the main auditorium forming a kind of Mohican haircut to the theatre. The alternating openings proved to not just not work but actually to put the whole system into reverse. The stack height was raised incrementally during the model tests until it exceeded the parapet level of the high-rise building nearby and much thought was put into the design of terminations that would be very robust to wind direction and turbulence.

The extract terminations

Various research papers were located that confirmed the efficacy of the H-pot and its more advanced form as a Cross-pot. These were imagined at a completely different scale to that at which the device normally manifests itself on domestic buildings (see Figure 15.5).

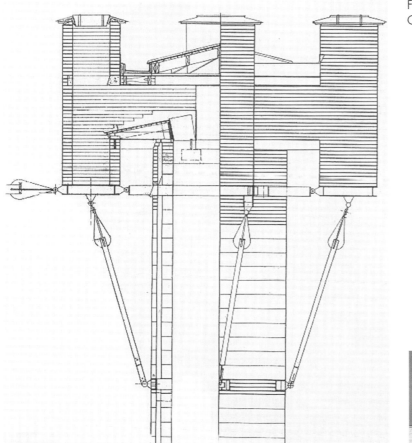

Figure 15.5
Construction drawing of masonry stack

Cross-pot/H-pot vents have been used to ventilate landfill sites and from the data derived from these experiments the final form of the building projects the stacks termination to the height of the surrounding buildings (3). The stacks themselves become silencers with arrays of acoustic vanes slotted within them. Masonry stacks are inevitably more effective and safer acoustically but steel framed stacks were configured on the existing lightweight roof of the main auditorium for structural reasons (see Figure 15.6).

15.9 The urban presence

The resulting forms are dramatic and distinguish each performance venue so that it is possible as one walks around the building to read it as a collection of different types of performance space. This is very different to the traditional form of an urban theatre building, where the distinct form of the

Figure 15.6
View up into a lightweight stack

auditorium tends to be lost in the poché of foyers and ancillary accommodation that bleeds out to the surrounding street pattern. Contemporary theatre buildings tend simply to continue this tradition, with some notable exceptions. One could speculate that the environmental strategy that treats each large space as a separate natural system tends to disaggregate the complex whole into its constituent parts (see Figure 15.7). Inevitably, the constituent parts can be rotated and shifted relative to each other so as to respond more locally to smaller-scale events in the urban landscape.

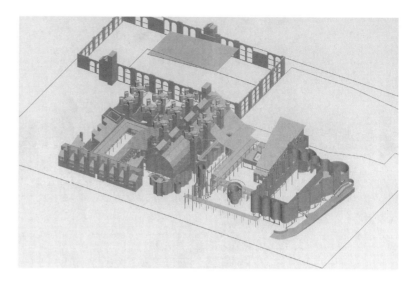

Figure 15.7
Contact writ large – the Coventry Arts and Media centre in Coventry City Centre

Figure 15.8
Foyer air intakes under construction

15.10 Encouraging air intake at pavement level

Designing viable low-level intakes in an urban environment is curiously problematic. Existing architectural vocabularies seem to be inadequate for the task. Prodigious free areas are required for a performance building and so the scheme naturally demanded much perforation at its base. We were anxious to avoid detaching the volume of the building from the surrounding pavement, the front line of the public realm, by a continuous line of louvered openings as one might find in a fully air-conditioned large building, or in many "comprehensively redeveloped" town centres venting basement car-parks. Air is introduced through arrays of domestic terracotta cavity ducts (see Figure 15.8).

Traditionally these are high-performing clay products that are embedded in walls and rarely seen, but they have a geometry that makes them well suited to becoming air intakes. The ducts are stacked up in a line of triangular buildings to force the maximum free surface area to supply air into the main foyers.

The rehearsal room employs the same device, admitting fresh air through heater boxes with a movable damper either connecting the room directly to the exterior or recirculating the air within, depending on the temperature and information about the levels of CO_2 received by the building-management system from sensors in each occupied space.

15.11 Synchronising cultural and environmental intent

The entrance elevation to the Contact (see Figure 15.3, centre right-hand side) is built up from cantilevered overlapping zinc screens that enclose the upper level bar and foyer. The patterns are derived from the embroidered hangings and banners that have been brought out for many generations to dress festivals in southern Mediterranean parishes. Our theory is that the shapes of the banners are derived from architectural moulding profiles supplied by local stonemasons. This is certainly the case in Malta where we found the stonemasons consulting Vignola's treatise to inform their stone-cutting technique. Fresh air is drawn up behind each projecting layer and taken through a simple heater box with a diverter, providing fresh air to that end of the foyers. The two vertical borders at either end clothe extract chambers.

15.12 Coventry University Library: the context in 1996

Whilst the Contact was under construction the practice won the opportunity to design a large (11,000 m²) library on an inner-city site. Coventry University is a post-1992 university developed out of Lanchester Polytechnic and associated colleges. The new publicly incorporated institution has had to develop its newly acquired university infrastructure very rapidly. One of the key components in this ambitious restructuring has been the construction of a Library and Learning Resource Centre, completed in August 2000 to replace the existing Lanchester Library (see Figure 15.9). The concept of the building is somewhat different to that of a traditional library in that it is an important teaching venue as well as being a repository for books and for scholars engaged in private study. It is densely provided with computers and the prognosis is for yet denser provision. A Learning Resource Centre contributes many of the broader student support requirements called for in contemporary university charters. It is not simply the new politically correct jargon for a traditional library. All of which is to say that the internal environment is potentially noisy, that students are allowed to talk, and that a certain amount of teaching happens on the principal floors in and amongst the IT provision and the books.

Figure 15.9
Aerial view of the new library

15.13 The client requirement for a deep plan

The Chief Librarian and his staff insisted on a deep-plan square building, the inverse of the type diagram that we had hitherto drawn, which attempts to dissolve a large programme into sequences of narrow sectioned elements coiled around each other. We wondered if it might be at all possible to make such a deep-plan building, 50 m by 50 m, naturally conditioned, which was also a much-emphasised ambition of the original brief. The Coventry inner ring road, elevated at this point, grazes the site, and a major interchange is adjacent at ground level. Furthermore, the need for security, and to discourage a latent temptation to throw books out of library windows for later collection, introduced the requirement to lock all the windows. We generated six alternative diagrams, all of which placed a large atrium in one part of the plan, a type plan of the last 30 years. The Institute of Energy and Sustainable Development at De Montfort University tested each plan type and we discovered that the introduction of a large atrium delivered a well-lit and reasonably well-ventilated but very narrow zone around the major void, and that the rest of the floor plate, apart from a narrow zone adjacent to the external elevations, was poorly served.

15.14 The development of a strategy to naturally condition a deep floor plate

It occurred to us that this large void could be redistributed to make fields of useful floor plate, alternating with smaller penetrations, so that one would redistribute and alternate this well-lit, well-conditioned perimeter zone. The more dimly lit spaces between became reasonable locations for book stacks. A simple and economical square grid, optimised at about 7 metres, set the rather crude rhythm of floor plate and void (see Figure 15.10). The next proposition was to draw fresh air beneath the principal floor and above the lower level through a continuous plenum, exposed to all four

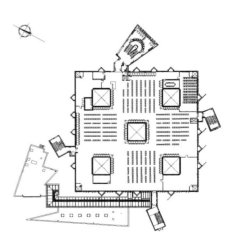

Figure 15.10
First-floor plan, Coventry

Figure 15.11
Section showing routes for air entry into the library

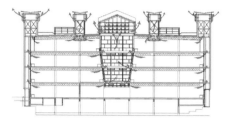

Figure 15.12
Section showing exhaust routes

Figure 15.13
Studying around an air-intake court

principal elevations, introducing fresh air into the building up through the four atria (see Figure 15.11).

This idea would deliver natural light and fresh air, warmed when necessary, to the readers and computer users and their tutors. The next proposition was to draw the warm air out through a large central atrium and perimeter stacks (see Figure 15.12). An effective and well-sealed damper was developed specifically for the project. A high level of local control is achieved by a fine grid of zones and close groupings of dampers. There are some 200 dampers in the building. This strategy provides a level of close, tight control, driven by the sophisticated building-management system considerably beyond that provided in our earlier buildings. As we worked through the logic of what is actually a very simple configuration, additional refinements were introduced. A glass lens was placed below the greenhouse tops of the intake atria to protect them from direct solar gain. A similar lens was placed at the base of the central atrium with a damper below soffit level, to choke the extract from the ground floor. The device enables library users to stand in the middle of the ground floor in the deepest part of the building and look up and see the sky. This has become a popular periodical-reading place (see Figure 15.13). The strategy seems to exclude traffic noise satisfactorily. If anything, the library floors are a little too quiet. There is no plant noise. There are no reports of the library being troublesome to clean. Particulate matter does not seem to be admitted into the building in any greater quantity than is expected by cleaners who attend to mechanically ventilated buildings.

15.15 The construction type and detail

There are no applied finishes to the walls or the concrete soffits and the ceiling heights are prodigious. The building assembles itself into a substantial urban *palazzo* and we pursued this analogy with enthusiasm. A contemporary illustration of the Palazzo Strozzi under construction became very useful to us; it shows how a principal elevation can arrive at a corner and fold back onto a much more utilitarian elevation, under which service rooms are stacked up at twice the rate of the principal floors. The Coventry north elevations double-bank inexpensive aluminium windows in a vague recollection of this principle. The 9 inch brick masonry modelled to the south-west and south-east elevations is in Flemish Bond and the principal window openings are protected by a *brise-soleil* embedded in the modelled masonry skin, formed out of cast-stone cills and brick piers. There is no need to make large frames of lightweight sun shading for these orientations at this latitude. They are the result of a particular set of aesthetic preferences.

15.16 The extract terminations

The rotated stacks between the window bays model the elevations at a coarser grain. The terminations to the stacks are taken well above obstructions on the roof and the converted motor-car factories nearby and develop into another type of termination. Split tubes of extruded aluminium are ranked, layered and displaced relative to one another to disrupt an incoming air stream whilst enabling a free discharge of air on the leeward side. We rather enjoyed the accidental evocation of a car radiator grille given that the building is in the heart of the declining Midlands motor industry. Fortunately the Coventry public, who retain a deep affection for motor cars, have taken to this imagery. The strategy is effective and we were interested to develop it. The opportunity came in the form of a commission to make a highly adaptable Learning and Teaching Building for Thames Valley University in 2001.

15.17 Thames Valley University New Learning, Teaching and Central Services Building: the context in late 2000

Thames Valley University is located in Ealing and Slough, just west of London, and is an amalgam of various former colleges. It is one of the principal exponents, with the City University, of the Government's intention to widen access to higher education in Britain. This is an extremely challenging task on the resource base allowed to new universities. It requires high levels of managerial agility and responsiveness as subject areas variously prosper or ebb away. The university's portfolio of curriculum offerings has been dramatically overhauled during the last few years and now it specialises in Health and Community Studies (funded by large Health Trust contracts), Business Studies, Media and Music. It inherited a large and valuable campus alongside the A4 motorway in the middle of Slough comprised of 1960s buildings suffering familiar construction inadequacies: the incorporation of asbestos, failing flat roofs and decaying lightweight cladding in the shadow of one of the principal landing and take-off flight paths to and from Heathrow.

The university's restructuring strategy proposes to dispose of a large component of the site and consolidate its learning and teaching accommodation requirements in a new building, with very high levels of inherent flexibility, on the residual site adjacent to the Paul Hamlyn Learning Resource Centre.

15.18 The strategy

This commission enabled us to develop the idea of a multi-storey, deep-plan flexible building and refine the original Coventry diagram. The Thames Valley building is broadly triangular and deploys three glazed penetrations, circular in plan, to extract stale air and deliver light deep into the building.

Fresh air is introduced from the perimeter, and from the centre through a device we have come to call the "fresh-air fountain" (see Figure 15.14). The fountain is fed by ducts at ground level and below the slab from each of the principal elevations. The budget is somewhat lower than that for the Coventry building. It has a simple steel frame, pre-cast concrete plank floors, a secondary steel frame along the perimeter with studwork infill, external insulation and, variously, render or cedar cladding.

The atria tops are developed into tiara-like semi-circular H-pot arrays (see Figures 15.15 and 15.16). All services are thrown outside of the principal plan area, which can be configured as completely or partly cellularised with large open-plan areas for administrators or computer suites, or as completely open floor plates. The south elevation filters solar gain seasonally through an external mesh sun-shade screen in which finer mesh panels trace the sunpath diagram, through the overheating months, onto the elevation behind. However, as higher education develops, it will require what they already describe as 24/7 accommodation. Much of its occupancy and use will be in the dark during the evenings with teaching predicted to happen until 9.00 p.m. The optimisation of the artificial-lighting scheme in the building will be as important as its ability to distribute good levels of natural light. Again, as at De Montfort, this new university is very interested in its own branding. Its students are busy people with full- and part-time jobs, making tremendous personal sacrifices to acquire higher education, many of them living in a nocturnal world at speed. A fascinating contrast in interests and priorities with the ancient universities, this institution has embraced the anticipated kudos of owning a new research level, green building.

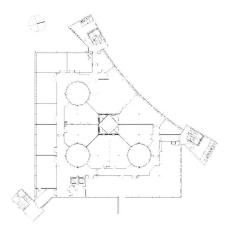

Figure 15.14
Thames Valley University first-floor plan

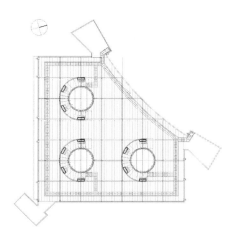

Figure 15.15
Thames Valley University roof plan

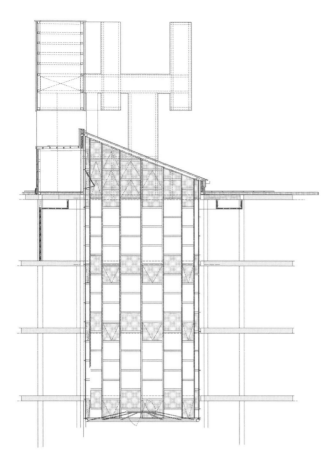

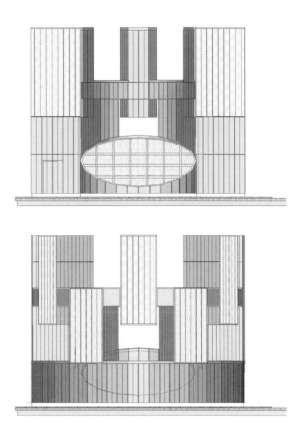

Figure 15.16
Detail of exhaust object, section and elevations

15.19 Conclusion

Throughout the last decade, our clients have coaxed us into modifying our rather rigid ideas of an optimal diagram for an environmentally responsible public-scaled building. By chance, they have all been in city centres, except for the first. We have grappled with demands for particular plan types, unpromising orientations, harsh acoustic and air-quality conditions. We have discovered that one cannot simply clip devices on to existing common building types; a greater contribution is required. We also suspect that "greenness", as an intent, aligns itself directly with a contextual intent in the broadest sense, and that fineness in detail and response offer the greatest rewards.

REFERENCES AND NOTES

1.   David Suchet to Stephen Daldry, and copied to Max Fordham, in a letter of 21 June 1996.

2.   Tim Lewers of Cambridge Architectural Research.

3.   Edwards, A, (1990), The Design and testing of wind-assisted passive landfill gas venting headworks. Project report, School of Physics and Materials, Lancaster University.

16

BedZED – Beddington Zero–fossil energy development

Bill Dunster

16.1   Introduction

Our definition of Zero-fossil Energy Development (ZED) is an excellent passive building envelope that reduces the demand for heat and power to the point where it becomes economically viable to use energy from renewable resources.

At BedZED, ZED has meant generating enough renewable energy over the course of a year to meet the whole of the new community's annual heat and power demands.

The ZED Challenge is to try to turn environmental problems into more sustainable urban solutions whilst working towards carbon-neutral new development in the UK.

The problems

A recent study (1) by Brenda and Robert Vale showed that an average UK family's annual carbon emissions were spent in the following way:

- one-third for heating and powering their home;
- one-third for private car use, commuting and land-based travel;
- one-third for foodmiles, with the average UK meal having travelled over 2,000 miles from farm to dinner plate.

This shows clearly that energy-efficient building design is important, but no more so than the other key factors determining everyday life in the UK today.

BedZED tries to show that it is possible to substantially reduce a typical household's carbon emissions, and hence reduce its overall ecological footprint at the same time as increasing overall quality of life. We found that the contemporary lifestyle in a conventional suburban home is currently so dysfunctional that it is possible to rethink each of these daily activities and reduce its carbon emissions while simultaneously providing quantifiable benefits to each individual.

This is important because we are no longer appealing to the niche green consumer but proposing alternative lifestyles that can be adopted by a wider market – but still making substantial progress towards creating a carbon-neutral urban infrastructure in the UK. We replace our urban fabric on average at around 1.5 per cent per year (2), showing that if ZED standards became commonplace, as a nation we could have weaned ourselves off our addiction to fossil fuels by the start of the next century, whilst still retaining historic buildings and city centres worth keeping.

However, much of our ordinary built fabric has little value and does not rise to the environmental challenges of the new millennium. Urban sprawl will cover 11 per cent of the surface area of the UK by 2016 (3). To avoid

Figure 16.1
The BedZED development

Figure 16.2
BedZED

Figure 16.3
Showhome bathroom

Figure 16.4
Showhome living-room

house prices rising, we are told we need an additional 3.8 million new homes by 2021 (4), not allowing for an increase in population – and it is hard for key workers such as teachers and nurses to find affordable homes in the South East. We are importing between 60 and 80 per cent of our food (5), and increased global competition for healthy organic food (currently 70 per cent imported) will become more intense as developing countries raise their expectations and as the world's population continues to rise. Competition for food, fossil fuels and water leads to conflict and results in the need to invest substantial military and trade resources to ensure supply. It would be helpful if a holistic accounting exercise was attempted that measured the actual financial and environmental implications of our current patterns of natural resource consumption. We also need to assess using the same criteria, the benefits of using local production to meet local needs.

16.2   Project overview

BedZED is a new community of eighty-two homes, eighteen work/live units and 1,560 m² of workspace and community facilities for the Peabody Trust in the London Borough of Sutton. In putting together our designs and proposals for BedZED, the ZEDteam tried to reconcile the apparently conflicting demand of more sustainable building with a higher quality of life.

Urban form

How do we create a compact city that people want to live in?

Figure 16.5
Mews showing sky-gardens

Figure 16.6
Achieving density and amenity

1.   At BedZED almost every flat has a small land- or sky-garden and a double-glazed conservatory, integrating the two features most desired by many suburban households.

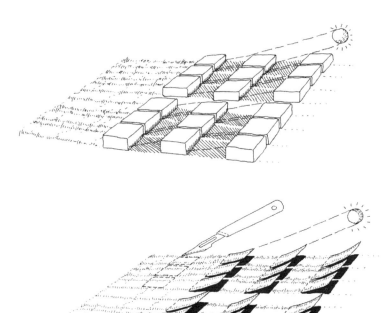

We hope to demonstrate that it is possible to reconcile the suburban garden village with the Urban Task Force's (6) agenda to substantially increase residential densities.

How do we reconcile higher densities with more amenities?

2. The central block of four terraces at BedZED (see Figure 16.7) achieves residential densities of over 100 homes per ha not including live/work units. This provides approximately 400 habitable rooms per ha and 200 jobs per ha at the same time as providing 26 m² of private garden/home, and 8 m² of public outdoor space/home. Even allowing for the entire BedZED site, including playing field, parking, CHP plant and community buildings, around fifty homes per ha is achieved. If these standards were commonly adopted it would be possible to reduce UK urban sprawl to about 25 per cent of its current footprint over the next century. This means that almost all of the new homes could be provided by brownfield-site regeneration, saving valuable agricultural land and green belt for biodiversity, leisure and locally produced organic food.

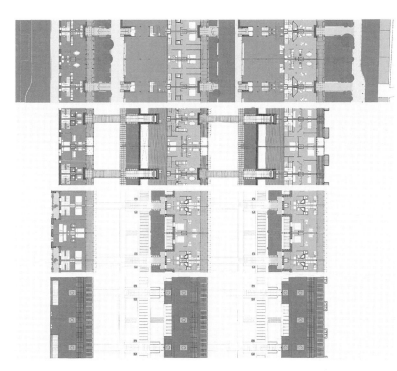

Figure 16.7
Slice plans showing ground level to roof

Energy

How do we run this new infrastructure off renewable energy sources?

3. It becomes possible to meet the new community's energy needs from renewable-energy sources by reducing the need for heat to around 10 per cent of a typical 1995 Building Regulations home. We do this by concentrating on maximising passive design measures. BedZED incorporates 300 mm minimum superinsulation, triple glazing, south-facing glazed sunspaces, thermally massive floors and walls, good daylight and passive-stack ventilation with heat recovery, and has energy efficient lighting and the latest "class A" white goods. By combining workspace with housing, the community's energy demand matches CHP output – often a problem with residential schemes.

4. A 135 kW electric output combined heat and power plant runs off 850 tonnes of woodchip per year at 30 per cent moisture content. This requires an area of short-rotation coppice woodland of around 70 ha, with around 24 ha of coppice being cut annually to meet the

Figure 16.8
Building physics diagram

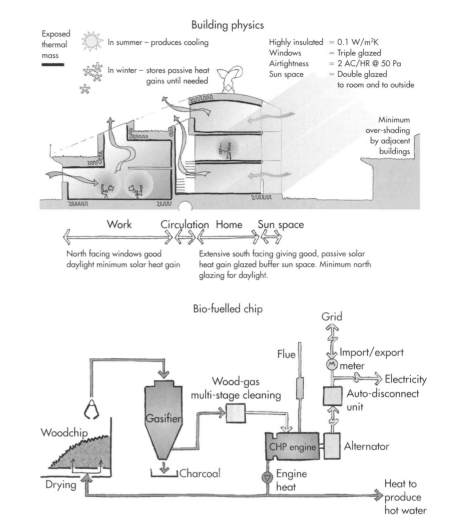

Building physics

Exposed thermal mass

In summer – produces cooling

In winter – stores passive heat gains until needed

Highly insulated = 0.1 W/m²K
Windows = Triple glazed
Airtightness = 2 AC/HR @ 50 Pa
Sun space = Double glazed to room and to outside

Minimum over-shading by adjacent buildings

Work   Circulation Home   Sun space

North facing windows good daylight minimum solar heat gain

Extensive south facing giving good, passive solar heat gain glazed buffer sun space. Minimum north glazing for daylight.

Figure 16.9
CHP diagram

Bio-fuelled chip

Grid

Flue   Import/export meter

Wood-gas multi-stage cleaning   Electricity
Auto-disconnect unit

Woodchip   Gasifier   CHP engine   Alternator

Charcoal   Engine heat

Drying   Heat to produce hot water

Figure 16.10
CHP woodchip store

community's fuel demand. At BedZED the woodchip is being sourced from the BioRegional urban tree station, which takes urban tree waste from the London Boroughs of Sutton and Croydon that would otherwise go to landfill. Initial research by BioRegional suggests that the embodied energy of the CHP plant itself is around 145,000 kWh, with equivalent carbon emissions of around 75,000 kg of CO_2. The CHP will produce around 726,000 kWh of electrical energy per year. If this energy were coming from a power station, it would be generating 326,000 kg of CO_2. As the CHP uses a renewable energy source, the CHP is displacing 326,000 kg CO_2 per year from the national-grid electricity production.

Transport

5. (See Figures 16.11 and 16.12.) A 109 kW peak solar-electric (photovoltaic) panel installation provides enough electricity to operate forty small electric cars, each running 13,700 km/year. Energy consumption of the electric cars varies according to make and size from about 20 to 30 kWh/100 km. A car pool is proposed on site to provide a mix of fossil-fuelled and electric vehicles. Since continental experience has shown that one pool car replaces four to five privately owned vehicles, the number of parking spaces can be reduced on site, increasing garden areas. Because of the currently high tax on fossil fuel, and the improved overall energy efficiency of the latest battery-powered electric vehicles, the photovoltaic-installation payback period is reduced from around 75 years when used to displace mains electricity to around 15 years when used to charge electric cars.

Figure 16.11
PVs on sunspace windows

How do we make green lifestyles more convenient than conventional alternatives?

6.   A farm shop selling locally produced, seasonal organic food is planned on site. Solar-electric-powered vans will collect a fresh food box from within their current range of around 40 km as a return trip, and 80 km if quick-charge facilities are available at the farmyard or countryside depot. Internet deliveries could further reduce the need for private car use.

How do we reduce the ecological footprint of the new construction to sustainable limits?

7.   Around 220 workspaces per ha at 12 m² per workspace have been achieved in the BedZED inner-urban block. With 309 bedspaces per ha, and given a few years for the community to mature, there should be enough employment opportunities on site for most residents to give up the daily commute to central London. In time, we hope to demonstrate that reverse commuting is taking place, where people living more centrally commute out to BedZED on empty trains at rush hour. This type of reverse commuting could double the existing capacity of the suburban lines without any additional infrastructure investment.

Figure 16.12
PV-powered electric van

Materials

How do we move towards sustaining the community from within its geographical bioregion (a natural region with distinct geological formations, climatic conditions and ecology)?

8.   As many of the bulky materials as possible have been sourced from within 55 km of the site, with only high-value items requiring special fabrication processes being sourced from outside the bioregion. No new materials were purchased unless the team was unable to source reclaimed material of the correct specification to the required programme, within the agreed cost plan. Almost the entire workspace steel frame was reclaimed, and all internal timber partitions were fabricated from reclaimed material. The construction site at BedZED almost entirely used traditional build techniques, slightly modified to match the requirements of the engineer's building-physics strategy.

Figure 16.13
Reclaimed timber

Building design

How do we visually integrate these new lifestyles within our existing urban fabric?

9.   There has been much architectural debate on the visual effect south-facing terraces might have on our existing urban landscape. At BedZED we have composed the gable ends to create a new type of street elevation (see Figure 16.14) and we propose this type of solar urbanism to be as valid as any more conventional urban strategy, where buildings are designed simply to face the street.

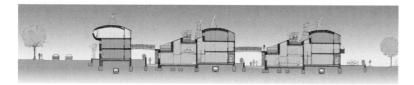

Figure 16.14
Section showing view of BedZED from London Road

Care has been taken to provide shelter and variety on the potentially defensive north elevations, where the need to minimise heat loss tends to reduce window size.

How do we socially integrate this lifestyle within existing communities?

10. BedZED effectively combines an office park with a housing estate at the same time as integrating communal sports and leisure facilities. This doubling up of land use offers potential developers more income than would a conventional single-use approach whilst respecting the normal planning restrictions on storey heights. Potentially this additional revenue could fund the increased costs of the carbon-neutral specification, enabling carbon trading to be integrated as part of the planning-approval process. This is exciting because local communities can potentially use the democratic UK planning process to increase the financial yield from sites within their jurisdiction and allow environmentally benign development to take place without relying on central government carbon-emission legislation – which will always be held back by the volume housebuilders lobbying power and resistance to change.

Figure 16.15
Funding carbon-neutral development

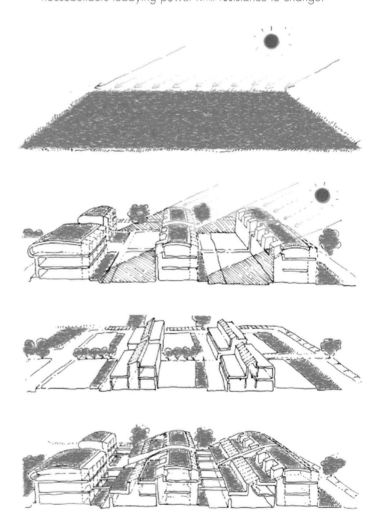

16.3 Development history

Starting as a theoretical exercise to redefine the garden-city concept over 6 years ago, BedZED is a response to the enormous demand for housing that the Government is predicting for the next 10 years and is one of the largest carbon-neutral developments of its type. While working at Michael Hopkins

& Partners, I was able to invite innovative consultants who were expert in workplace design to apply some of the latest thinking to housing. They aided in the creation of Hopetown, designed for an Architectural Association symposium on urban sustainability in 1996. The theory was converted into practice in the form of my own home, Hope House in East Molesey, Surrey, completed in 1996. The lessons learned from both were transferred directly to BedZED. Bill Dunster Architects (BDA) then teamed up with BioRegional, an independent environmental organisation specialising in creating sustainable environments through the use of local resources, and the Peabody Trust, a client able to understand, analyse and manage innovation without being frightened of it. Together they sought a site for the next stage. Once found, the team bid against direct commercial competitors and won, despite not being the highest bidder.

The development of this site in Sutton is significant for several strategic reasons.

- The location is in one of the few London boroughs that, at the time, was actively promoting Local Agenda 21 issues.
- The site is in close proximity to the largest area of green open space in south London. The adjacent open land is to be developed into an ecology park.

Figure 16.16
Green space surrounding BedZED

Site

- The site was relatively low value and the London Borough of Sutton's (LBS's) Unitary Development Plan (UDP) had already identified the need for an ultra-low-environmental-impact housing development. An independent environmental-impact assessment by Aspinalls, commissioned by LBS, suggested that a discount of around £200,000 could be awarded in favour of the Peabody Trust, setting a useful precedent for local authorities seeking "best value" when disposing of land.
- The proximity to Hackbridge station, good local bus routes and the new tram-line connection between Wimbledon and Croydon make it easier to reduce car dependence on this site.

Using planning gain to finance carbon-trading initiatives

The outline planning brief gave permission for 305 habitable rooms on the site and this density limit created the market value of the site. The current scheme achieves 271 habitable rooms with the addition of 2,369 m² of commercial space, which provides in excess of 2,000 m² of income-generating accommodation without the Trust actually paying for the land this space is built on. This planning gain will fund the increased building costs of an ultra-low-energy, super-insulated build specification. The domestic energy consumption should be around one-tenth of that of a standard suburban home built to 1995 building regulations.

16.4 Urban design

The rectangular site shape and Sutton Council's suggested access point from the busy London Road effectively bisected the new development, creating a dense central block of six south-facing terraces with both a vehicle and a pedestrian mews. Non-allocated parking is flung to the perimeter of the site, making the central island a pedestrian- and cycle-prioritised homezone. Terraces are never more than six-homes long, allowing a central pathway to run through the centre of the new live/work development, gradually expanding in width to create a village square overlooking a new sports pitch.

Figure 16.17
BedZED site plan

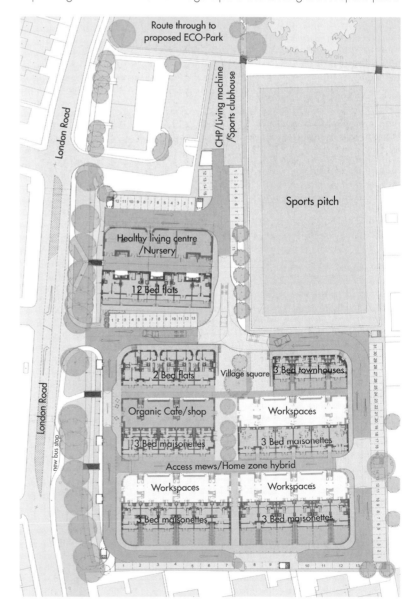

It is hoped that long-term plans will see the football pitch in turn opening out onto a constructed wetland designed by the Wildfowl and Wetland Trust – turning the surrounding metropolitan open land (formerly sewage-treatment land) designated for gravel extraction into a useful wildlife sanctuary and future ecology park. An existing services conduit supplying energy to the existing housing estate to the south-east bisected the northernmost block, separating the workspace from the housing. On most of the other terraces, workspace is placed in the shade zone of housing (see Figure 16.17).

The new homes are always higher than the workspace and have raised gardens, allowing residents to overlook workers – rather than the other way round! The new terraces have been carefully sculpted to maximise sunlight penetration to as many homes as possible for as much of the year as possible – testing the principles of solar urbanism. The spaces between the blocks have been carefully composed to maximise access to private defensible open space, land-gardens and sky-gardens at the same time as minimising overlooking from the work spaces and maximising internal daylight levels. The new elevation to the London Road resembles a series of waves crashing towards the south (see Figure 16.18), composed from gable ends, articulated to express the differences between housing and workspace. Roundbacks provide shelter to the raised access causeways on the north elevations of the central block, softening the northern elevations with their defensive, small windows and bringing useful increases in floor area to the topmost rooms.

Figure 16.18
London Road wave-like elevation

The combined heat and power (CHP) plant has been integrated with the on-site grey- and black-water treatment and the sports clubhouse and changing rooms – showing how these facilities can have minimal landtake in compact urban developments.

The environmental masterplan: the evolution of the Solar Urban Village

The sectional design of the buildings allows the roof surfaces of the workspace to become gardens for the adjacent housing (see Figures 16.19 and 16.20). This approach enables a relatively high occupation density (in excess of 500 people/hectare for the developed area of the site) and also provides good access to sunlight and high levels of amenity for all the residents. A 45 × 90 m football pitch and a small village green provide open space for children, and will provide useful sports facilities to the surrounding community. An integrated transportation strategy promoting public transport, pedestrians and cyclists, together with an on-site taxi/car pool available to both businesses and residents, should substantially reduce the emissions normally caused by private car use. Internet bulk food delivery from local supermarkets further reduces shopping trips – and a planned telecommuting business centre offering video conferencing and access to expensive equipment normally found in central London offices will be encouraged on site.

The project attempts to set up a repeatable urban block, with good solar access on a flat, typically urban site, so that one building does not steal

Figure 16.19
Sky-gardens

Figure 16.20
Bridge to sky-gardens

Figure 16.21
Sunspace PVs

sunlight from its neighbours. Four different urban typologies have been designed at BedZED:

1. south-facing terraces with front gardens;
2. terraces with workspace and vehicle mews;
3. pedestrian-only mews with gable end entry workspace and café/shop;
4. covered arcade with housing/nursery and health clinic.

These four building-blocks, each with different subdivisions creating a variety of flat types, could enable the BedZED system to be replicated on other sites.

The BedZED scheme approaches the highest density of naturally lit mixed-use urban grain capable of benefiting from useful amounts of passive solar gain and retaining the opportunity for photovoltaics whilst providing each household with an outdoor space. Early studies on CHP sizing undertaken by Ove Arup Partners (OAP) showed that passive solar gain contributed up to 30 per cent of a typical home's annual space-heating requirements, although some homes at BedZED are more shaded by adjacent building-blocks than others. Any new development can become carbon neutral if enough biomass fuel can be sourced to power a combined heat and power plant. However, biomass will soon become a valued renewable-energy source, and in a carbon-neutral future it will become important to make limited fuel stocks power the maximum number of new homes and workplaces. The UK currently only produces around 2 per cent of its power needs from renewable-energy sources. Even if the British public universally adopted green power tariffs from renewable-energy schemes, only a tiny fraction of potential demand could be met by renewable sources. Each new offshore windfarm built will be required to power our existing historic towns and city centres that cannot be so easily upgraded to minimise energy consumption. Seen in this context, the necessity of minimising energy demand through passive design becomes one of the most important key-form-generating criteria, and perhaps a good reason why new urban fabric should maximise solar access to new homes.

Design precedents and research initiatives

BDA have co-ordinated the following research initiatives, feeding the results back into design development. Working with the same team responsible for two EU Joule studies on ultra-low-energy workspace, the New Parliament Building Westminster and Nottingham University New Campus, the latest environmental research has informed the detail design of BedZED. The team visited both the new campus buildings and a super-insulated earth-sheltered housing terrace at Hockerton, demonstrating the theoretical viability through built examples. The work of Brenda and Robert Vale at the autonomous house in Southwell and at the Hockerton earth-sheltered community has inspired our own practice and at the same time set new performance standards by which we believe all future work in this field should be judged.

16.5 Environmental strategy

BedZED uses 300 mm of insulation everywhere (see Figure 16.22), substantially reducing the heat loss through the building fabric. Most glazing is placed to receive maximum sunlight, with unheated double-glazed sunspaces being an integral part of each dwelling. Thermally massive concrete ceilings and tiled floors provide enough heat storage to prevent overheating in summer and to retain warmth in winter to compensate for overcast days. In summer, enough windows open to turn each glass

Figure 16.22
Super-insulated walls

sunspace into an open-air balcony, avoiding the need for expensive external shades. A wind-driven passive ventilation system provides around 60 per cent efficient heat recovery with no electrical energy consumption (see Figure 16.23), and low-energy domestic white goods and lights reduce the load on the combined heat and power plant to around 150 kW for the whole site. The combination of passive solar gain, super-insulation and heat recovery has enabled heating systems to be removed from the dwellings, using heat lost from a centrally placed hot-water cylinder connected to the CHP. By placing workspace in the shade zones of the housing terraces, high IT users obtain cool daylight without heat, avoiding the need for energy-wasteful active cooling systems.

Combining workspace with housing provides a more consistent load for the power plant, which is entirely wood-fuelled. Trees absorb CO_2 as they grow and return it to the atmosphere when burnt – making BedZED one of the first high-density urban carbon-neutral developments in the UK for a century or so.

Local sourcing

Throughout the scheme we have tried to specify natural materials and products with a low embodied energy. Materials and products will be sourced, where possible, from within a 55 km radius of the site in order to reduce pollution from transportation and to encourage local industry. A default to a building-industry conventional product is only used when cost or programme make a local product or service unviable.

Figure 16.23
Wind-driven ventilation cowls

Recycled materials

Reclaimed pine doors, recycled second-hand steel beams and columns for the larger-span frames in the workspace units.

Local materials

Locally sourced WWF FSC timber oak, ash and various softwoods, local aggregate from the adjacent gravel pit for concrete, local bricks and blocks, locally sourced reclaimed steel, timber and aggregate etc.

Local services

Wherever possible, local businesses were used to supply labour to build BedZED. Although great efforts have been made trying to persuade local joinery firms to make up the good-quality oak-framed sunspace glazing on the south elevations, Danish windows were specified in the final event. On future projects we are examining the potential for an on-site covered joinery workshop making simple components such as straight-flight external staircases, landscape bollards, oak bridges and decking. Without local sourcing, the thermally massive construction (needed to provide the weekly thermal-storage cycles required to maximise the potential of winter passive solar gain) could have high-embodied-energy penalties.

16.6   Detail design

A 6.3 m crosswall spacing with clear-spanning precast slabs used throughout the project provides good flexibility to accommodate a variety of different unit types whilst reducing the number of floor-plank variations. Four different floor constructions were examined: tunnel-forming; *in situ* bison precast planks; and an Omnia permanent-shuttering system. The Hollow core Tarmac planks provided a fair face soffit – important for thermal storage – and the *in situ* topping allowed more flexibility in creating apertures and stairwells. All *in situ* concrete has been purchased from the nearest batching plant a few hundred metres away. Heavyweight concrete blocks from the Thames Estuary are used for crosswalls, with 300 mm cavities and two-part Ancon stainless-steel ties connecting an outer skin of local Cranleigh brick and local oak weather-boarding fixed to a softwood balloon frame. By keeping all brickwork in the same plane as areas of timber cladding it has been straightforward to change cladding materials on elevation whilst maintaining continuity of the super-insulated cavities. All north/east and west windows are triple-glazed with low E coatings, using a proven Scandinavian timber-frame system. The project size and timescale did not make it easy for the team to develop and test a local product – although this should become a priority for future jobs.

Searching for timber-framed triple-glazed rooflights has not been easy in the UK, and we have modified the well-proven Vitral aluminium system, with its minimal sight lines and ease of incorporating opening lights. Roof gardens containing 300 mm of soil are laid over a two-layer bituminous-felt flat-roof system – well proven and with a 20-year guarantee. Recycled-steel beams sourced from local demolition sites allow larger clear spans in the workspace, and reclaimed-softwood joists and floorboards were used to build the mezzanine floors. All internal staircases were made from locally sourced softwood. Although it was hoped that reclaimed-pine internal doors would be used throughout the project, that was not possible and it was necessary to buy in new WWF FSC-certified solid new doors from B&Q. It is believed that the reclaimed materials (when available) are cost-effective, remove the need to procure new material, and add significant value to prospective inhabitants.

Large water tanks normally used for motorway drains store rainwater from the uppermost roofs for flushing toilets, and water-saving spray taps and showers were specified throughout the project. Unitised bathroom pods were market tested but were not viable for the relatively small project size. Prefabricated unitised service walls incorporating bathroom fittings were investigated to assist the construction programme, but again these needed larger economies of scale. Wind-driven ventilation cowls (using standard easily cleanable heat-recovery units) use durable stainless steel running gear previously developed by Dunster's team at Hopkins for the Jubilee Campus at Nottingham University. All technologies used are proven and presented few problems to local contractors. Much design effort has been invested into reconciling traditional low-environmental-impact materials with the OAP building-physics concept, specifically avoiding thermal bridging and maintaining airtightness.

16.7   Integrated transport strategy

The largest contribution to minimising the carbon emissions from transportation is made by encouraging residents to telecommute or work on site. However, it is equally important to minimise private fossil-fuel-powered car use.

A car pool will be incorporated in the development and secure bicycle stores are provided for all dwellings and workspaces. The CHP has been sized to suit both the peak thermal requirements of the BedZED "village" and the annual electricity demand, to avoid the expensive and wasteful heat dumping that would be required if it were sized to meet the instantaneous peak electrical loads. A significant design challenge has been to match both electrical and thermal requirements on the CHP system, enabling it to run as efficiently as possible. For this reason, solar-electric photovoltaic panels have been integrated into the south-facing conservatory roofs to enable varying numbers of electric-powered vehicles to run off zero-carbon-emissions renewable-energy sources without oversizing the CHP plant.

The development is technically capable of requiring no fossil fuel for its inhabitants' lifestyle needs within the geographical constraints of the London Borough of Sutton – although this is dependent on each individual's personal adoption of the communal facilities integrated within the development. For journeys where walking or cycling is impractical, the best-developed solution for zero-carbon-emission transport at present is that of electric vehicles, where the electricity can be generated from a renewable source. With a range of around 50 miles, electric vehicles are suitable for the short trips that make up more than 95 per cent of car usage. Petrol cars are typically at best around 20 per cent efficient (petrol calorific content to delivered axle power) and worse than this in urban running. Electric cars are around 70 to 85 per cent efficient, partly because of the inherent efficiency of the electric motor and also because of the energy-saving techniques such as regenerative braking. Even allowing for the typical power station, transmission and charging losses, the electric car's carbon emissions are lower than those of the petrol-powered car. If the electricity is generated from a renewable source such as sunlight then the CO_2 emissions drop to zero.

The typical annual distance travelled by a family car is 16,000–20,000 km (27–34 miles a day). An efficient electric vehicle such as the Citroen Berlingo van will travel around 6.2 km per kWh of electrical energy. For an average annual distance of 15,000 km (quite heavy use bearing in mind that the vehicle will not be used for very long journeys), this would require 2,420 kWh annually.

The fuel costs are also lower for the electric car as a result of the high tax level on petrol. Typical costs for a petrol-driven car are 4.9 p/km, while an electric car would cost around 1.3 p/km if run on mains electricity (not including road tax, insurance, servicing, etc., which would be significantly cheaper for an electric vehicle because of reduced complexity, increased reliability and increased longevity). The current lifetime of the batteries (typically Nickel Cadmium, though now being superseded by Nickel Metal Hydride) before replacement or reconditioning is required is around 5 years (or 80,000 km). Options for battery hire and recycling by the car manufacturer already exist.

Currently the payback period for PVs to displace fossil-fuel use in building is around 73 years. However, the BedZED analysis shows that if PVs are used to reduce the fossil fuel used for transport, the payback period can be reduced by a factor of five, to about 13 years.

Bill Dunster Architects has run a modified Citroen Berlingo van, operating from a combined live/work unit at Hope House, for three years with no problems. The van's annual electric requirements are met by a 1.35 kW peak BP Saturn solar array, integrated into the sunspace roof.

At the BedZED development we are trying to persuade Citroen to make the Hope House van modifications to people carriers at the factory, and set up a financial package with BP Solar enabling occupants to lease-purchase the photovoltaic cells with the car. We have incorporated enough PV to run forty electric vehicles, each running around 13,700 km/year, of which around thirty could be privately owned and ten could be part of a communal car pool that could include some petrol cars for longer journeys. Charge points are already inexpensively integrated within street-light lampposts. Zero-emission solar powered electric vans would be the ideal vehicle for collecting and distributing locally produced organic food.

The external circulation at the BedZED scheme is designed to give priority to pedestrians and cyclists. Cars and other vehicles are accommodated for but

Figure 16.24
Sunspace glazing

Figure 16.25
Mechanical and electrical systems

must defer to other users. Following this philosophy we have adapted the "Home Zone" approach to the specifics of the Beddington site (7).

Building integrated renewable-energy features

The PV installation is glass or glass mono-crystalline BP solar modules and is integrated into a double-glazed low-E conservatory roof and vertical softwood windows. The rooflight modules are inclined at around 30 degrees and replace sheets of clear glass at the same time as providing an element of fixed shading. Rainwater run-off from the PV surfaces is collected in underground tanks and used to flush toilets. Heat recovery from the rear surface of the PVs in the form of warm air collecting at the apex of each conservatory is achieved simply by opening the patio doors inside each flat. Extensive work was needed with the window manufacturer to integrate electric-cable routes within a relatively low-embodied-energy standard product.

Each home has an unheated passive solar sunspace, capable of providing up to 30 per cent of a typical home's annual space-heating requirements, depending on the extent of overshadowing from surrounding buildings.

The woodchip-fuelled combined heat and power plant has been totally integrated into one building, housing both the "Living Machine" grey- and black-water treatment plant and the sports clubhouse and changing rooms (see Figure 16.25). The land take required by the CHP is minimal for a development of this size.

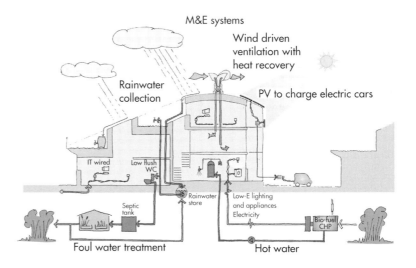

M&E systems

16.8 Conclusion

Funding the green specification

Perhaps the most important message is that the wholehearted approach to sustainable design is more cost-effective than simply tacking on a few green initiatives to a standard-volume housebuilder's product. The quantity surveyors for the project have calculated that the same amount of money per unit is simply spent in a different way, allocating more funds for super-insulated walls and glazing at the same time as reducing the central-heating budget to two finned-tube radiators in each home. Apply this logic to the whole build and barriers to sustainable development begin to disappear.

Future ZEDs

BedZED has been chosen as a Housing Forum demonstration project, and we are currently working with the Housing Forum on sustainability indicators (see scenarios 1 and 2 in Table 16.1). We entered the recent Peabody Trust/Architecture Foundation "Living in the City" competition and have begun work that shows how the BedZED system can be prefabricated using a flat-pack semi-volumetric approach. BDA with BioRegional are currently working with all members of the existing design team on this new initiative, called ZEDfactory. Operating from shared offices in one of the workspace units of the completed project, ZEDfactory will provide off-the-shelf zero-fossil-energy development solutions using tried and tested components – effectively breeding carbon-neutral development from the BedZED parent.

We are now developing the BedZED system as a standard house-type product, with a supply chain integral and the perceived higher-risk, "green" technological elements already priced and performance tested. Therefore we can source local materials, suppliers and contractors for the structural frame and fabric of the base build, but still have the benefit of the specialist environmental modelling and innovative components that would ordinarily be unaffordable. But, most importantly, we can provide planners and local councils with detailed environmental-performance projections before even a

Table 16.1
Densities of scenarios 1 and 2

	Scenario 1 BedZED actual	Scenario 2 BedZED inner city
Total area of site	16,544 m² (1.65 hectares)	6,397 m² (0.64 hectares)
Total number of dwellings	82 homes	63 homes
Total floor area in m²	10,388 m² (gross external)	8,235 m² (gross external)
Area of sports pitch	4,336 m²	–
Area of village square	538 m²	538 m²
Sports club and CHP buildings	497 m²	–
Area private gardens at ground level/roof	2,058 m²	1,638 m²
Area of primary circulation routes	3,207 m²	1,160 m²
Vehicular and pedestrian mews	540 m²	540 m²
Area of car-parking	986 m²	zero
Total no. of residents	244	198
Total no. of habitable rooms	271	225
Total no. of workers (1 person per 12 m²)	196	130
Total no. of car-parking spaces	84	–
Area of work space	1,695 m² gross external	1,695 m² gross external
	1,559 m² gross internal	1,559 m² gross internal
Area of commercial space	871 m² gross external	–
	810 m² gross internal	–
Summary		
Residential population density/hectare	148/hectare	309/hectare
Working population density/hectare	119/hectare	203/hectare
Total site population density/hectare	267/hectare	512/hectare
Habitable rooms/hectare	164/hectare	352/hectare
Parking spaces/hectare	51/hectare	–
Public open space/hectare	2,954 m²/hectare	840 m²/hectare
Private gardens and sky-gardens/hectare	1,247 m²/hectare	2,560 m²/hectare
Total green space (6 + 7)/hectare	4,200 m²/hectare	–
Homes/ha @ 3.5 hab. rooms/home	47 homes/hectare	100 homes/hectare
Annual carbon emissions to atmosphere	zero	–
No gas supply to site	–	–
Installed capacity of renewable energy harvesting/m² of floor area	47 W/m² peak	–
Installed capacity of biomass chp per m² floor area	21 W/m² thermal, 14 W/m² electric	–

site has been purchased. This enables the environmental performance to be traded as planning gain within Section 106 agreements (part of the Town and Country Planning Act 1990, which regulates development or use of land). It is hoped that these initiatives will substantially remove many of the barriers to change that are currently obstructing large-scale carbon-neutral developments of this type in the UK. We hope to show that these benefits create added value to the developer that will exceed the additional construction costs, and we believe that in time the market will simply vote with its feet and request ZED developments in preference to conventional alternatives.

Figure 16.26
ZEDfactory logo

ZEDteam members

Client	The Peabody Trust
Environmental Consultants	BioRegional Development Group
Architects	Bill Dunster Architects
Mechanical and Electrical Engineers, and Building Physicists	Ove Arup Partners
Structural and Civil Engineers	Ellis & Moore
Quantity Surveyors	Gardiner & Theobald

REFERENCES

1. Vale, B. and R. (2000), The New Autonomous House. Thames and Hudson, London, p. 210.

2. DLTR website (then DETR): www.dltr.gov.uk or www.statistics.gov.uk

3. As reference 2.

4. The Urban Task Force (1999), Towards an Urban Renaissance: final report of the Urban Task Force. E&FN Spon, London, p. 46.

5. DETR General Information Report No. 53: Building a Sustainable Future, and SAFE Alliance Foodmiles campaign 1998.

6. As reference 4.

7. Information on Home Zones is available at: www.homezonenews.org.uk

BoO1: an ecological city district in Malmö, Sweden

Michael Sillén

17.1 Introduction

From Industrial wasteland to a leading ecological area

BoO1 – City of Tomorrow – is an entirely new, ecological district of 9 ha for 600 dwellings as well as offices, shops and other services. It is the first development stage of Västra Hamnen (The Western Harbour), one of Malmö's growth areas of the future. The area is typical of the redundant, contaminated industrial urban land that has affected the environment, but it also has many positive aspects such as its location by the sea, the beach and the city centre. When fully developed, the densities are expected to be 122 persons/ha, 216 habitable rooms/ha and 72 dwellings/ha.

A fundamental ecological approach to planning, building and construction is a key tool in the creation of the district. Innovative ideas and new techniques will enhance the environmental standard of the area.

The aim is for the district to be an internationally leading example of environmental adaptation of a densely built urban environment. It will also be a driving force in Malmö's development towards sustainability.

Figure 17.1
The Western Harbour area in Malmö from above

17.2 Design principles

Man's interaction with the environment is a fundamental factor in the planning of the district. The district is designed to be ecologically and socially sustainable. Access to green areas and water, utilisation of daylight, and varied visual and auditory impressions create an environment in which people feel well.

Substances that can damage the environment or human health are avoided in the construction of the houses. The buildings are designed so that those who live and work in them can do so with minimal environmental impact and resource use. When the buildings are eventually demolished the construction materials should be recyclable.

In order to realise the environmental ambitions a quality programme has been worked out. Figure 17.2 shows the mixed-use, environmentally sensitive, high-quality architecture. The programme defines the standards that must be met by those who participate in the construction and building process. The quality programme is directly linked to the agreement on the granting of land between the developers and the city of Malmö. This means that the requirements of the programme are accepted by the developers and are part of their commitment to each building project.

Ecologically harmful substances are banned from use in construction. Materials that impair the quality of sewage or sludge are prohibited where there is a risk of leakage to the water supply and sewage system. Materials and techniques that facilitate reuse and recycling at the end of a building's life are being employed.

Figure 17.2
Architecture in BoO1

Information technology is used to make it easier for people to live their daily lives comfortably and in ways adapted to the environment. Monitoring of water and energy consumption will be possible in the dwellings. The exact departure time for buses will be indicated, and the car belonging to a car pool of the district can be booked on the home computers. The residents are continually supplied with information on the results of refuse sorting and other environmental matters.

Two particularly energy-efficient detached houses will be built in the district. In these two model houses, which will also become display apartments during the fair, the best commercially available techniques as regards to energy requirements and a sound indoor climate will be applied. The aim is to demonstrate the feasibility of attractive houses with very low energy requirements at a reasonable cost. The environmental measures complement the architecture much more than is normally the case. The developers employ well-known Swedish and international architects to guarantee high architectural quality.

17.3 Energy

One hundred per cent of the district's energy is from renewable sources, generated in or near the area. Sun, wind and water are the basis, together with energy from refuse and sewage.

A large percentage of the heating is extracted from the sea and aquifers (natural water storage in the bedrock). Heat will also be produced by solar collectors (see Figure 17.3).

Electricity is mainly generated by wind power and to a lesser extent by photovoltaics.

a

b

Figure 17.3
Solar collectors
a. Vacuum solar collectors on roof
b. Solar collectors

Biogas is extracted from refuse and sewage from the area and is returned after purifying to the district via the city's natural gas system.

The solar collectors and photovoltaic systems, including those in private properties, are operated and managed by Sydkraft in order to ensure high maintenance and operation standards.

The requirement for 100 per cent renewable energy means that there must be a balance between production and energy use on an annual basis. It is intended that in the future all energy used in the area will be produced there.

However, at present the new electricity grid and district heating network is linked to the existing systems of the city in order to bridge the time-lapse between the point of production and use of energy, without the need for specialised equipment for energy storage. The city's system is used as a store and as a backup supply.

Efficient and minimal energy use is essential in order to reach the target of entirely locally produced renewable energy. The buildings in the district are designed to reduce the demand for heat and electricity. The target for average energy use on the properties is not to exceed 105 kWh per square metre gross room area annually (typical figures are 56 and 43 kWh/m²y for heating and electricity, respectively). This includes all energy related to the property; heating, hot water as well as electricity for households and for running the building services. Household equipment, lighting and other electric installations are the most energy-efficient on the market.

CO_2 emissions are zero as the energy is from renewable sources. Emissions from traffic are marginal in the area and there are no emissions from industry there.

To minimise heat losses from the houses it is important to reduce the thermal transmittance of the buildings. Generally this is done by increasing the thermal insulation of the buildings and by installing energy-efficient triple-glazed windows with low-emissivity coatings.

17.4 Water and waste

Household waste is expected to be 325 kg/person/y. 53 per cent of this is organic and will be treated at the biogas digester; 22 per cent is mixed waste, which will be incinerated; 25 per cent will be recycled.

Eco-cyclic adaptation of the water and sewage system is based on the co-ordination of initiatives in the properties in the district with those throughout the rest of the city. The area is connected to Malmö's existing sewage system, which is currently being improved.

Vacuum refuse chutes that extract organic waste from other household rubbish are installed in the district to deal with 60 per cent of total household refuse (see Figure 17.4). Disposal hatches outside the properties lead to holding tanks from which the waste makes it way to two docking stations at the edge of the area.

The two fractions are either transported to the biogas digester or to the waste-incineration plant. Recyclable packaging materials are collected at special points close to the houses.

The installation of kitchen-sink waste mills in some of the area's homes increases the separation of organic waste and thereby the potential for

Figure 17.4
Separation of waste in a vacuum refuse chute

biogas extraction. There are separate drains for kitchen waste, leading to a shared separation tank.

A new biogas digester is being built at the sewage-treatment plant to treat the organic waste from the area. The biogas plant converts the organic waste to fertiliser and biogas for heating and vehicle fuel. The organic waste comes from waste water, waste mills and from the vacuum refuse chutes.

New facilities to extract nutrients and heavy metals from the sludge at the treatment plant are under construction. The plant will separate phosphates from the sludge so that they can be returned in the form of essential nutrients to agricultural land. This avoids the risk of contamination by sewage sludge and also means a lower transport requirement. Metals used in the treatment process will be reused. The remaining solid sludge will then be burned as a bio-fuel. Another advantage of this technology is the reduced need for chemicals at the water-treatment plant.

The concept is based on new ideas and to some extent new technology and has not been fully developed anywhere else in the world.

Water use is estimated to be less than 200 litre/person/day.

17.5 Traffic

The area has been planned for environmentally responsible transport. Dense development with a wide array of services and recreational options will reduce the residents' needs to travel out of the neighbourhood.

Residents and companies in the area have access to mobility-management information (see below). As well as environmentally friendly vehicles, the bicycle circulation system may be the most important element of the green traffic strategy. The network of walkways and bicycle paths is being built to a high standard to ensure that it is perceived in all situations as a very attractive choice for short trips.

Public transport within the district is a vital component of the strategy. In order to attract passengers, the bus system will be comprehensive from the outset so that it becomes the logical choice for residents. All bus lines connect to the more important destinations in Malmö.

Significant investment is being made in vehicles powered by environmentally sensitive fuels. Public-transport vehicles run on such fuels, with the transport fleet made up of electric- and gas-powered and hybrid vehicles (see Figure 17.5). The vehicles used to maintain the area will eventually all be electric-powered.

To ensure that environmentally sensitive vehicles can be easily used, the neighbourhood features a filling station for natural gas and quick charging of electric vehicles. Slow charging of electric cars is possible in special parking spaces.

"Green" vehicles are given priority access to the city district and to parking. The intention is to progress towards all traffic in the city district running on renewable fuels.

Residents of the city district are invited to join a car pool made up of green vehicles. Studies show that time-share ownership of cars leads to less-frequent and better-planned car trips.

A Mobility Office has been opened in the city district. The Mobility Office gives residents and businesses advice and information about

Figure 17.5
Environmentally friendly vehicles

environmentally sensitive transport, and carries out various programmes aimed at changing transport behaviour.

17.6 Biodiversity and the green environment

The structure of green space has a central role in the creation of an environmentally sustainable city district. Our aim is to create a green and pleasant area that optimises biodiversity, despite the high density of building.

A rich variety of habitats with a long-term sustainable biodiversity is created in parks and gardens.

Certain habitats are created in the back yards whilst the parks are primarily used for habitat types that do not fit with or are not appropriate for gardens. Half of the yard area is planted. The district also includes two parks – the Canal Park and the Strand Park.

In order to have as much vegetation as possible, a green-point factor has been applied. This system means the building contractors compensate for the area they have built on by providing green spaces. Example of green spaces that give points are plant beds, foliage on walls (creepers and climbing plants), green roofs (grass and mostly moss-stonecrop sedum carpets) (see Figure 17.6), water surfaces in ponds and large trees and bushes.

Figure 17.6
Habitats in public spaces

Figure 17.7
Green roofs

Green points

From a list of thirty-five "points", the building contractors have selected at least ten points. Example of green points that benefit biodiversity are:

- a bird-nest box for each apartment;
- a bat box for each plot;
- a courtyard containing a traditional cottage garden with various sections;
- a part of the garden left to grow by natural succession;
- a courtyard containing at least fifty Swedish wild flowers.

Other points bring landscape architectonic qualities to the courtyard, whilst still others facilitate rainwater run-off processing.

Rainwater is dealt with locally (see Figure 17.8) without any connection system and without using holding areas other than the sea and the canals. Rainwater is cleaned and treated through a surface run-off system, which makes only minor demands on the drainage system.

17.7 IT

Information technology is used as an active tool to improve the environmental performance of the area and also to facilitate a lower-energy, more environmentally friendly day-to-day life for the residents.

IT is used to measure, control and regulate different sub-systems and there are opportunities for individual charging for the residents' energy use. Residents also have the opportunity to follow their own and the district's energy use.

Road-information-technology systems are being used to inform the public and to control traffic in the area. Public-transport vehicles are let through first

Figure 17.8
Local treatment of rainwater

at intersections controlled by traffic lights. Bus shelters feature real-time information displays that show when buses will arrive. Residents are also able to tap into the city district's broadband network to look up arrival times for buses and book cars from the car pool from the comfort of their flats.

Opportunities for remote work from home and e-commerce also reduce the residents' needs to travel out of the area.

Different initiatives are needed to meet the communication and documentation needs of the district. An environmental web TV channel is a less-conventional form of communication, which is used to inform, engage and influence the public.

FURTHER READING

Hancock, C. (2001), Malmö: towards the sustainable city. EcoTech 4, November, pp. 40–45.

CONTACT DETAILS AND WEBSITES

Tor Fossum
City of Malmö
Environmental Department
SE – 205 80 Malmö
Telephone: +46 40 35 95 68
www.ekostaden.com

Bo01 – City of Tomorrow: www.bo01.com

Appendix A

Solar energy, temperature and photovoltaics

A.1   Solar data

As can be seen from Figure A.1, annual irradiation is similar in much of northern Europe (sometimes referred to poetically as "the cloudy North").

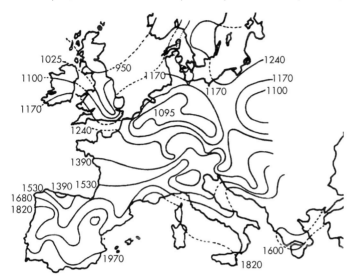

Figure A.1
Solar irradiation over Europe
(kWh/m²/y) (1)

Figure A.2 shows the pattern of the sun's movement across the sky and Table A.1 gives data for a lattitude of 52°N.

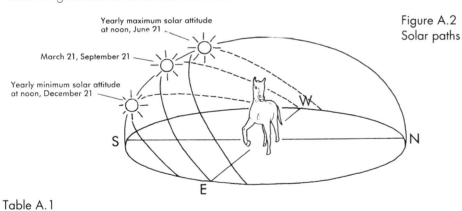

Figure A.2
Solar paths

Table A.1
Approximate solar altitudes and azimuths at 52°N (2)

Date and time	Altitude (degrees)	Azimuth (degrees)
21 December		
09.00	5	139
12.00	15	180
15.00	5	221
21 March and 22 September		
08.00	18	114
12.00	38	180
16.00	18	246
21 June		
08.00	37	98
12.00	62	180
16.00	37	262

A.2 Temperature data

Figure A.3(a) gives air and ground temperatures at Falmouth in 1994 (3) and Figure A.3(b) gives air temperatures on a warm day (10 July 1994) in Garston, north of London.

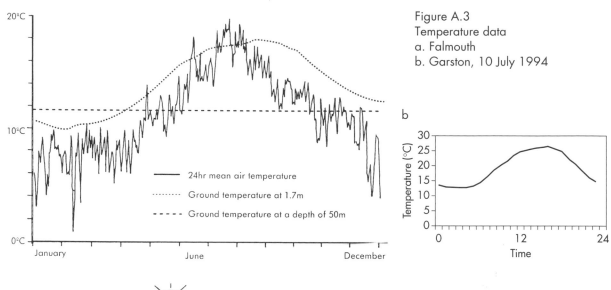

a

24hr mean air temperature

.... Ground temperature at 1.7m

--- --- Ground temperature at a depth of 50m

Figure A.3
Temperature data
a. Falmouth
b. Garston, 10 July 1994

A.3 Photovoltaics

PV systems convert solar radiation into electricity. They are not to be confused with solar panels, which use the sun's energy to heat water (or air) for water and space heating.

The most common PV devices at present are based on silicon. When they are exposed to the sun, direct current (d.c.) flows as shown in Figure A.4. PVs respond to both direct and diffuse radiation (see Figure A.5) and their output increases with increasing sunshine or, more technically, irradiance.

Common PVs available are monocrystalline silicon, polycrystalline silicon, thin-film silicon and thick-film silicon. A typical crystalline cell might be 100×100 mm. Cells are combined to form modules.

Table A.2 shows a number of current different types of PV cell and their approximate efficiencies.

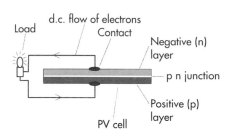

Figure A.4
Diagram of PV principle

Higher efficiencies – of over 30 per cent in some cases – are being achieved using multilayered structures. New materials such as copper indium diselenide (CIS) are being investigated and work is also under way on the use of dye-sensitised solar cells and organic films.

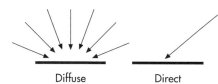

Diffuse   Direct

Figure A.5
Direct and diffuse radiation

It is also useful to keep efficiencies in perspective. A tree (see Figure A.6) relies on photosynthesis, a process that has been functioning in seed plants for over 100,000,000 years and only converts 0.5–1.5 per cent of the absorbed light into chemical energy.

Figure A.6
A Cambridge tree, near an array of seventeenth century solar collectors (i.e., windows)

Table A.2
PV efficiencies (4, 5)

Type	Approximate module efficiency
1. Monocrystalline silicon	12–15
2. Polycrystalline silicon	11–14
3. Thick-film silicon (polycrystalline silicon)	9
4. Thin-film silicon (using amorphous silicon)	5

More recently, the national grid has proved only 25–30 per cent efficient in providing us with electricity from fossil fuels.

Crystalline silicon cells consist of p-type and n-type silicon and electrical contacts as shown schematically in Figure A.4. The cells, which are of low voltage, are joined in series to form a module of a higher, more useful voltage. Figure A.7 shows one of many ways of building in PVs.

Modules electrically connected together in series are often referred to as a string, and a group of connected strings as an array. An array is also a generic term for any grouping of modules connected in series and/or parallel. Power from the array (see Figure A.8) goes to a Power Conditioning Unit (PCU), which converts the electrical output from the PV array into a suitable form for the building. The a.c. output from the PCU goes to a distribution board in the building or to the grid if supply exceeds demand.

How much energy do PV systems produce?

The output from building-integrated PV installations is the output of the PV array less the losses in the rest of the system. The output from the array will depend on:

- the daily variation due to the rotation of the earth and the seasonal one (due to the orientation of the earth's axis and the movement of the earth about the sun);
- location, i.e., the solar radiation available at the site;
- tilt (see Figure A.9);
- azimuth, i.e., orientation with respect to due south (see Figure A.9);
- shadowing;
- temperature.

A rule of thumb for maximum annual output is that the optimal azimuth is due south and the optimum tilt is the latitude minus 20° but considerable choice is possible around this 100 per cent point (see Figure A.10).

Very roughly, a monocrystalline silicon PV array will produce about 100 kWh/m²/y in London with an azimuth of due south and a tilt of 30°.

For comparison, Table A.3 gives the output of a number of different 50 m² arrays.

Figure A.7
Typical module construction
(glass/EVA/Tedlar™/Polyester/
Tedlar™)

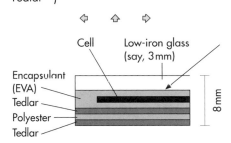

Figure A.8
Schematic of a typical grid-connected PV system

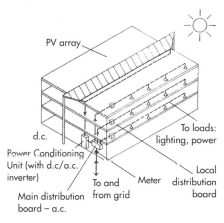

Table A.3
Comparison of array outputs (MWh/y) (London data; unshaded arrays)

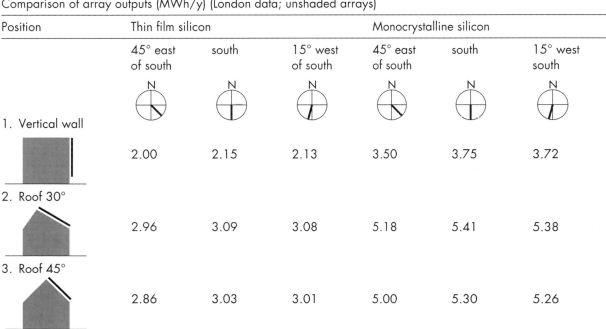

Position	Thin film silicon			Monocrystalline silicon		
	45° east of south	south	15° west of south	45° east of south	south	15° west south
1. Vertical wall	2.00	2.15	2.13	3.50	3.75	3.72
2. Roof 30°	2.96	3.09	3.08	5.18	5.41	5.38
3. Roof 45°	2.86	3.03	3.01	5.00	5.30	5.26

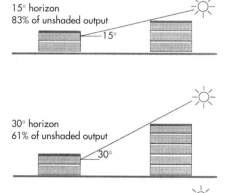

In urban situations overshadowing can be a key issue. Figure A.11 shows a loss of output due to (continuous strips of) neighbouring buildings.

Potential of a site for photovoltaics

The use of a site's potential for PVs can be evaluated quantitatively. A very simple starting point is to examine the orientation and angle of the roofs and a figure corresponding to the annual solar irradiation of the site.

Thus, for example, we may examine the roofs indicated on the housing development illustrated below, a truncated form of Parkmount, transplanted to London (see Figure A.12).

The amount of solar radiation (expressed as a percentage of the maximum for an ideally sited roof) is determined by using the yearly irradiation map for London (see Figure A.10). Maps such as this exist for a variety of locations.

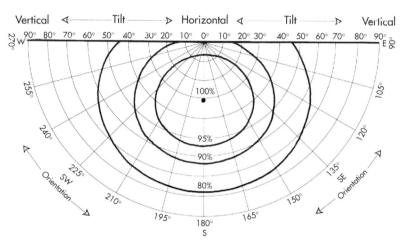

100% corresponds to the tilt and orientation that gives the maximum total annual solar radiation (1,045 kW/m²/y on a surface oriented due south at a tilt of 31°) on a fixed surface in London (51°36′N, 0°03′W)

Figure A.10
Yearly irradiation map for London

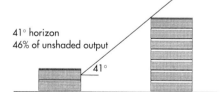

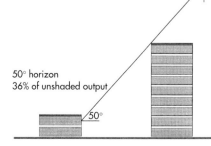

15° horizon
83% of unshaded output

30° horizon
61% of unshaded output

41° horizon
46% of unshaded output

50° horizon
36% of unshaded output

Figure A.11
Shading effects by neighbouring buildings

A figure (which may be called the Index of Solar Utilisation) can then be derived using the roof areas and their percentage of solar radiation:

Roof	Area (m²)	Orientation (° from north)	Tilt (° from horizontal)	Solar radiation (% of maximum)
1	86	196	15	96
2	86	196	15	96
3	86	189	15	96
4	86	189	15	96
5	86	183	15	97

For the roofs in Figure A.12:

Index of Solar Utilisation

$$= \frac{(86 \times 0.96) + (86 \times 0.96) + (86 \times 0.96) + (86 \times 0.96) + (86 \times 0.97)}{430}$$

$$= 0.96$$

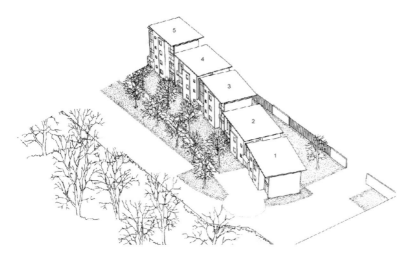

Figure A.12
Roof analysis

REFERENCES

1. Thomas, R. (2001), Photovoltaics and Architecture. Spon, London.

2. Thomas, R. (1999), Environmental Design. Spon, London.

3. Bunn, R. (1998), Ground Coupling Explained. Building Services, December, pp. 22–27.

4. As reference 1.

5. Data from Solar Century: www.solarcentury.co.uk

Appendix B

Wind and water

B.1   Wind

Wind energy

Figure B.1 shows the energy in the wind in the UK at a height of 10 m in open countryside (1); the units are gigajoules (GJ)/m²/y. Since 1 GJ = 278 kWh, 3 GJ = 834 kWh and we see that this is close to the amount of solar radiation falling on a horizontal surface in London (see Appendix A).

Wind turbines

Wind specialists are familiar with the well-known Betz equation:

$$P = 0.645 \, (A \times V3)$$

Where P is the power an ideal wind turbine can extract from the wind in kW.

A is the swept area of the turbine in m² (i.e., the area within the shape formed by the blades as they rotate, thus, for a horizontal axis machine this would be the circle described by the blades), and V is the wind speed in m/s.

Aerodynamic, mechanical and electrical losses in real machines are likely to reduce this by one-third (2).

The advantages and disadvantages of different turbines are given in Table B.1.

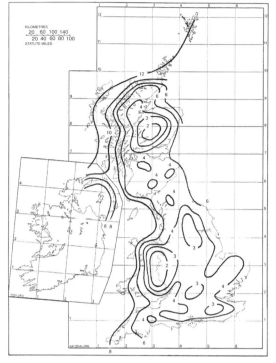

Figure B.1
Energy in the wind

Table B.1
Comparison of axis orientation for small-scale turbines (3)

Type	Status	Remarks	Some examples
Horizontal axis			
Upwind, lift	Large series, proven product	Widely used, mainly in open fields	Inclin, Aerocraft, Bergey, Vergnet, Lagerwey
Downwind, lift	Proven product	Mainly in smaller power ranges, also for open fields; building integration example	Proven
Vertical axis			
Savonius, drag	Proven product	Generally very silent and reliable, storm resistant, low efficiency	Windside, Shield's Jaspira turbine (GTi1)
Darrieus, lift	Small series	Several subtypes exist, often as prototypes, simple construction, fair efficiency; noise and vibration still to be covered	
Combined lift and drag	Prototypes	Simple construction, no external start up needed	Globuan, Solavent
Large turbines	Small series	Low-noise, reliable	AES

B.2 Water

Figure B.2 shows a schematic geological section with an aquifer and Figure B.3 shows aquifers in the south-east of England.

Figure B.2
Schematic geological section (4)

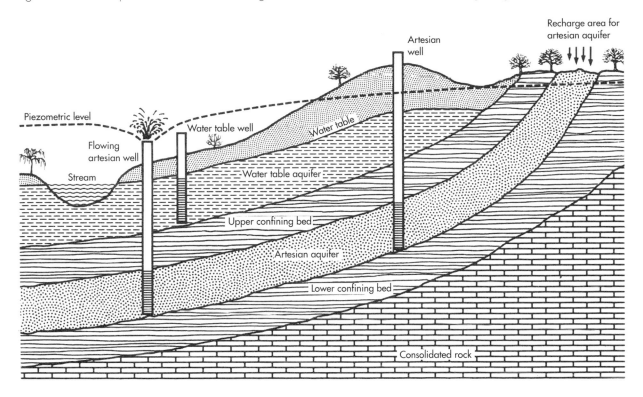

Figure B.2
Schematic geological section (4)

REFERENCES

1. Rayment, R. (1976), Wind energy in the UK. BRE CP 59/76. BRE, Garston.

2. Thomas, R. (1999), Environmental Design. E&FN Spon, London.

3. Timmers, G. (2001), Wind Energy Comes to Town: Small Wind Turbines in the Urban Environment. Renewable Energy World, May–June, pp.113–119.

4. Anon. (1985), Water-to-water heat pumps. Technical Information EC 4708/8.85, The Electricity Council, London.

5. Cooper, E. (1990), Utilisation of Groundwater in England and Wales. Water and Sewerage, pp.51–54.

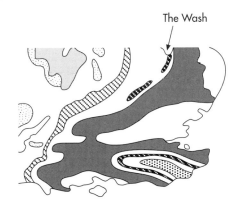

Legend

Major aquifers

Chalk
Perma-triassic sandstones

Minor aquifers

Hastings beds
Lower greensand
Jurassic limestone
Magnesian limestone
Carboniferous

Figure B.3
Aquifers in the south-east of England (5)

Appendix C

Air quality

Air pollution damages health, buildings and the environment.

Pollution can be characterised by a long-range component, an intermediate-range component and a "local" component.

In the urban environment traffic pollution can account for up to 85 per cent of total emissions compared to a national average of 25 per cent (1). Pollutants resulting from traffic include CO, oxides of nitrogen (NO_X), CO_2, hydrocarbons and particles.

The UK Government has focused on the eight pollutants shown in Table C.1 in its strategy for ensuring that "everyone can enjoy a level of ambient air quality in public places which poses no significant risk to health or quality of life" (2). (The comments in the table are based on a variety of sources.)

Pollutant levels in cities will depend on background levels and local effects, which can be greatly influenced by urban form and the pattern of winds sweeping through the city. Solar radiation will enter areas at different times and will change the pollutant mix. There will be significant variation with the contaminant studied and generalisation is difficult. Figure C.1 shows NO_2 concentrations in a street canyon in London (7).

Table C.1
Air contaminants

Key air contaminants	Comments
1. Benzene	Main source is combustion and distribution of petrol (i)
2. 1,3 Butadiene	Note 1; major source is road transport
3. CO	Main source is road transport and, particularly, petrol vehicles (4). Strongly dependent on local traffic speeds and congestion conditions
4. Lead	Arises from petrol use – use of unleaded petrol is reducing the level
5. NO_2	(i)
6. Ozone	(ii)
7. Particles PM_{10}	(iii)
8. SO_2	Principal source is fossil-fuelled power stations; contaminant is then dispersed by wind so elevated sources are dominant (iv); difficult to map.

Notes

(i) There is an effect of "roadside enhancement" for benzene, 1,3 Butadiene and NO_2 (3). All combustion processes in air produce oxides of nitrogen (NO_X). In polluted areas such as heavily congested towns and cities nitric oxide (NO) concentrations can exceed those of NO_2. Engineers should specify low-NO_X boilers.

(ii) "Ozone is a transboundary pollutant – concentrations depend on emissions in the rest of Europe as well as the UK" (5).

(iii) PM_{10}: there are three categories of sources:
primary emitted by combustion processes;
secondary formed in atmosphere from chemical reactions;
tertiary so-called coarse particles formed from non-combustion sources – wind-blown dusts soils, fires, tyre debris, etc. Volcanoes also naturally produce a great deal of dust.

(iv) (6).

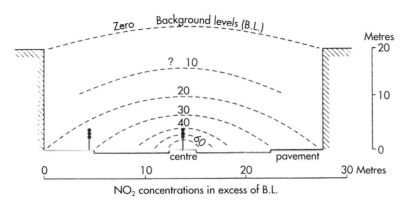

NO$_2$ concentrations in excess of B.L.

Figure C.1
NO$_2$ concentrations in a London street canyon

REFERENCES

1. Liddament, M. W. (1997), External Pollution – the Effect on Indoor Air Quality. In Rooley, R. (ed.), Indoor Air Quality and the Workplace. MidCareer College, Cambridge, UK.

2. Anon. (2000), The Air Quality Strategy for England, Scotland, Wales and Northern Ireland. London, DETR.

3. As reference 2, p. A102.

4. As reference 2, p. A126.

5. As reference 2, p. A159.

6. As reference 2, p. A102.

7. Laxen, D. P. H. et al. (1987) Nitrogen dioxide distribution in street canyons. Atmos. Environ. 21 (8), 1989–1903. Cited in Boucher, K. (1990/91), The Monitoring of Air Pollutants in Athens with Particular Reference to Nitrogen Dioxide. Energy and Buildings, 15–16, pp. 637–645.

Appendix D

Acoustics

D1   Urban noise levels

Traffic is the main source of noise in the city now, just as when Joyce described two barmaids listening to horses in the streets of Dublin with the words "Bronze by gold heard the hoofirons, steelyrining imperthnthn thnthnthn" (1).

Broadly, what one sees is that where the noise level is high in a city there should be a strategy for sound attenuation.

Table D.1
Typical urban noise levels

Location	Noise level dBA	
1.	A quiet living room in the evening with the windows closed	26–28
2.	Background noise levels in central London at night	39–49
3.	A small urban park with traffic and church bells in the distance and small birds nearby	46–47
4.	A street with only occasional traffic	49–51
5.	Busy offices	45–55
6.	Lively urban area during the day	65–75
7.	A very busy street with cars, buses and the occasional motorbike	75–85
8.	Tube train decelerating into a station	80–84
9.	Tube train accelerating out of a station	84–87

REFERENCE

1.   Joyce, J. (1961 edition), Ulysses. Random House, New York, p. 256.

FURTHER READING

Peliza, S. (1994), Noise and Natural Ventilation. Building Services, June, pp. 49–50.

Appendix E

Fuel cells, turbines and engines

There are several ways of providing combined heat and power (CHP) that are under development: fuel cells, microturbines and Stirling engines. All will move the generation of electricity closer to the point of use and in the case of the smallest units, directly into the home.

E1 Fuel cells

Fuel cells are silent and vibration-free (both important for buildings) and produce no or very-low levels of pollution. They can be made in a variety of sizes. In them the chemical energy of the fuel is converted to thermal energy and electrical energy directly. Because this process is not limited by the Carnot efficiency (see Glossary), fuel cells can achieve higher electrical efficiencies than conventional combustion processes. Table E.1 shows some characteristics of the principal types.

E2 Microturbines

"Small" gas turbines are of variable size. Commercially available units include those that produce several kW (electrical) and others that produce

Table E.1
Fuel-cell types and characteristics (1,2)

Item	Fuel-cell type	Operating temperature °C	Fuel	Electrical efficiency % now/(target)	Comments
1.	PEM (proton exchange membrane)	30–690	Hydrogen	35/(45)	Units operating at 80°C could be used for vehicles or for CHP in buildings; "small" units could provide 200 kWe – micro might provide 5 kWe; high-quality H_2 required
2.	PAFC (phosphoric acid fuel cell)	220	Hydrogen	<42	Well-developed technology; suitable for CHP in larger buildings; in use at Woking; tolerates a less-pure H_2 than for PEMs; some run on biogas; typical output: 200 kWe, 200 kW thermal
3.	AFC (alkaline fuel cell)	80–200	Pure hydrogen	40–60	Mainly used in spacecraft; occasional terrestrial applications
4.	MCFC (molten carbonate fuel cell)	650	H_2, CO, CH_4, others	47/(60)	Suitable for 200 kW–2 MW systems; CHP and stand-alone systems
5.	Solid oxide fuel cells	700–1000	H_2, CO, CH_4, others	47/(65)	Suitable for CHP systems and for combined cycle systems (which combine fuel cells with turbines); 2–1000 kW range; CHP and stand-alone systems, a unit that produces 1 kW of electricity and 3 kW of heat is being tested (see Chapter 6)

300 kWe (3). They are undergoing intensive development for use both for CHP and standby electrical generation.

Recently, a unit with an output of 45 kWe and a modulating thermal output of 80–255 kW was installed in a tower block with seventy-two homes in London; it was selected in part because of the comparatively low level of CO_2 and NO_x emissions compared with reciprocating-engine CHP (4).

E3   Stirling engines

The engine patented in 1816 by the eponymous Reverend Dr Robert Stirling alternately heats and cools a gas in a confined volume with part of the engine space being kept hot by an external heater. Advances in piston technology are overcoming some of the previous difficulties experienced.

Stirling engines are potentially much less polluting than internal combustion engines. Units with an output range of 1–20 kW (electrical) are commercially available, are undergoing tests or are on the drawing boards.

REFERENCES

1.   Lehmann, A. K., Russell, A. and Hoogers, G. (2001), Fuel cell power from biogas. Renewable Energy World, November–December, pp. 76–85.

2.   Larminie, J. (2000), Fuel cells. Ingenia, 1(4), pp. 43–47.

3.   Stephens, M. (2002), Diesel still rules the standby power world. Electrical Review, 22 January, pp. 18–19.

4.   Anon. (2002), Microturbine CHP set for London district heating. eibi (energy buildings and industry), January, p. 20.

Appendix F

Landscape

It is agreed that vegetation has numerous beneficial effects as shown in Figure F.1. Conditions can be improved for both the users of the public space and those in the buildings surrounding it.

But does vegetation really make a difference? Well, yes, but how much so will depend very much on location, urban form, species, density and countless other factors. None the less, let us follow others and be intrepid in putting some numbers to a few items.

Photosynthesis can be represented by the equation

$$CO_2 + H_2O \xrightarrow{\text{Sunlight}} CH_2O \text{ (Carbohydrate)} + O_2$$

F1 Plants

Figure 1.6 gave an estimate of CO_2 uptake for a mixture of lawn and trees – uptake will depend greatly on choice of vegetation and local conditions. Photosynthesis, of course, requires light; at night plants will respire and produce CO_2.

The loss of water by plants, known often as evapotranspiration, keeps the plants cool, reduces air temperatures and increases the humidity. The exact effect on temperature in urban neighbourhoods depends very much on context. A study in Greece found that as a rule of thumb, a 0.8 K reduction in temperature resulted from a 10 per cent increase in the ratio of green to built area [1].

Trees can remove airborne particles by trapping them until they are washed away by rain. The extent of the reduction will obviously depend on the nature of the vegetation. An early reference [2] suggests a 75 per cent reduction in particulates from a densely planted area of perhaps 15 m in height and 30 × 30 m in plan.

Vegetation can also reduce noise levels. For example, a tree-filled park of a minimum width of 30 m will provide 7–11 dB attenuation (125–8,000 Hz) [3].

The useful effects of vegetation as a wind-break are shown in Figure F.2.

In designing our cities we should, of course, ensure that the trees don't actually interfere significantly with passive solar gain. Figure F.3 shows that a horse chestnut in December can still block out a significant amount of solar energy. Figure F.4 gives data for a number of North American barren trees.

Finally, the table below classifies a selection of plants under the categories of being native to northern and central Europe, pollution resistance and attraction to birds or insects. Most perform well under more than one category.

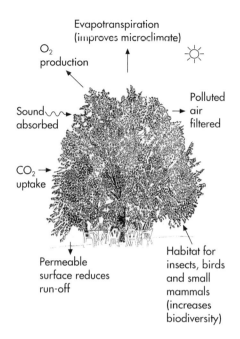

Figure F.1
Beneficial effects of vegetation

Figure F.2
Reduction in wind resulting from a
good shelterbelt (4)

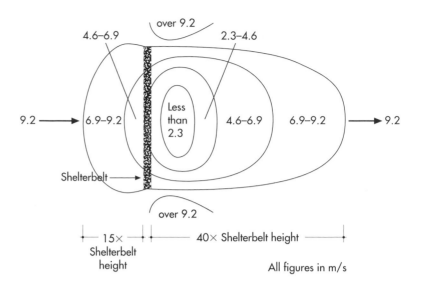

Figure F.3
A horse chestnut tree in Cambridge in
July and December (5)

Table F.1
Plant characteristics

Plant name	Approx. height (in m)	Native species	Pollution resistance	Attracts birds	Attracts insects
Deciduous trees					
Acer platanoides "Clumnare"	20–25	✓	✓		✓
Betula pendula	8–22	✓			✓
Carpinus betulus	5–15	✓		✓	✓
Castanea sativa	10–35	✓	✓		✓
Eyonymus europaeus	2–6	✓	✓	✓	✓
Fagus sylvatica	25–30	✓			✓
Fraxinus excelsior	20–35	✓	✓		✓
Ilex aquifolium	2–5	✓	✓	✓	
Ligustrum vulgare	2–5	✓	✓	✓	✓
Malus sylvestris	5–10	✓		✓	✓
Populus tremula	10–15	✓	✓		✓
Prunus padus	5–10	✓	✓	✓	✓
Prunus spinosa	1–5	✓	✓	✓	✓
Quercus petraea	20–35	✓	✓	✓	✓
Quercus robur	30–35	✓		✓	✓
Salix alba	10–20	✓	✓		✓
Sambucus nigra	2–7	✓	✓		✓
Sorbus aucuparia	5–10	✓			✓
Tilia cordata	25–30	✓	✓		✓
Viburnum lantana	2–5	✓	✓		✓
Coniferous trees					
Juniperus communis	0.5–4	✓	✓	✓	
Pinus sylvestris	10–30	✓		✓	✓
Taxus baccata	8–15	✓	✓	✓	✓
Thuja occidentalis	15–20		✓	✓	

Figure F.4
Relative light penetrations of various
barren trees (heights indicated are
approximations) (6)

Young ash
5 m

Native pecan
14 m

Sassafrass
8 m

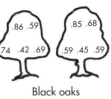
Black oaks
10 and 12 m

REFERENCES

1. Dimoudi, A. and Nikolopoulou, M. (2000), Vegetation in the Urban Environment: microclimatic analysis and benefits. PRECis: Assessing the Potential for Renewable Energy in Cities. Centre for Renewable Energy Sources, Pikermi, Greece.

2. Fitch, J. M. (1975), American Building: The Environmental Forces That Shape It. Schocken, New York.

3. Cited in Rayden, D. (2000), State of the Art on Environmental Urban Design and Planning. In PRECis: Assessing the Potential for Renewable Energy in Cities, Project Coordination: The Martin Centre, University of Cambridge.

4. Anon. (1964), The Farmer's Weather, Ministry of Agriculture, Fisheries and Food, Bulletin No. 165. HMSO, London.

5. Littler, J. and Thomas, R. (1984), Design with Energy: The Conservation and Use of Energy in Buildings. Cambridge University Press, Cambridge.

6. Holzberlein, T. M. (1979), "Don't let the trees make a monkey of you". Proceedings of the Fourth National Passive Solar Conference. Newark, Delaware: ISES – American Section.

Appendix G

Materials

Table G.1
Broad comparative energy requirements of building materials (1)

Material	Primary energy requirement (GJ/tonne)	
	Worldwide	UK
Very-high-energy		
Aluminium	200–250	97
Plastics	50–100	162
Copper	100+	54
Stainless steel	100+	75
High-energy		
Steel	30–60	48–50
Lead, zinc	25+	
Glass	12–25	33
Cement	5–8	8
Plasterboard	8–10	3
Medium-energy		
Lime	3–5	
Clay bricks and tiles	2–7	2–3
Gypsum plaster	1–4	
Concrete:		
In situ	0.8–1.5	1.2
Blocks	0.8–3.5	
Precast	1.5–8	
Sand-lime bricks	0.8–1.2	
Timber	0.1–5	0.7
Low-energy		
Sand, aggregate	<0.5	0.1
Flyash, volcanic ash	<0.5	
Soil	<0.5	

REFERENCE

1.   Thomas, R. (1999), Environmental Design. Spon, London.

Appendix H

A simplified assessment method

H1   SUSAM-E: Simplified Urban Sustainability Assessment Method – Environmental

The disclaimer is that this is a simplistic, naïve, reductivist approach, which conveys almost nothing of the richness of the city and may in fact be pernicious if used incorrectly. "To predicate sustainable practice exclusively on quantifiable indicators is over-simplistic, for the true measures of sustainability are also qualitative and cultural" (G. Haworth, Chapter 14). None the less, it is hoped that this may be of some help as a first step (particularly for students) in assessing an urban project from an environmental viewpoint.

Obviously, social and economic aspects are also of great importance and, perhaps, similar simplistic assessments will be developed, with questions such as whether the project allows for social inclusion, has mixed use, reaches the critical mass for provision of facilities, provides easy access to all, encourages job opportunities, and so on.

	Points	Your assessment
1. Does the urban form contribute to sustainability and if so how?	10	
2. Is energy relating to transport reduced and if so how?	10	
3. Are the building designs and grouping of the buildings energy efficient and is the solar potential thoroughly considered? Give a quantitative explanation of this. This could take the form of a solar utilisation index, i.e., the percentage of the roof area that would provide, say, 95 per cent of the maximum PV output (see Appendix A).	10	
4. Does landscape enhance the environment and if so how? (NB This could include the appearance of the neighbourhood or the microclimate around the buildings.)	10	
5. Explain how the energy demands of the building will be supplied in a sustainable way.	10	
6. Explain the contributions to sustainability of your solutions for choice of materials, water use and waste recycling.	10	
TOTAL		

Glossary

Acid Rain   Pollution caused principally by oxides of sulphur (SO_x) and nitrogen (NO_x) emitted during fossil-fuel combustion and metal smelting. This leads to formation in the atmosphere of acids, which are returned to earth.

Brownfield Land   Brownfield land is normally understood to mean land that has been previously built on but is now derelict, underused or of low ecological value.

Carnot Efficiency   The Carnot limit is the maximum efficiency for a heat engine, e.g. an internal combustion engine, and is given by: (TH – TC)/TH where TH is the temperature of the heat source and TC is the temperature of the heat sink in degrees Kelvin.

CFC   A chlorofluorocarbon. A source of depletion of the ozone layer and a contributor to global warming.

Dioxin   A name given to a number of chemicals formed at high temperatures (including in bonfires), some of which are carcinogenic.

Electrolysis   In the electrolysis of water, an electric current is passed through it to produce hydrogen and oxygen.

Electromagnetic Spectrum   The electromagnetic spectrum covers the full range of radiation from X-rays to microwaves. Solar radiation falls within this and consists of UV (290–400 nm), visible (400–760 nm) and infrared (760–2200 nm).

Energy
- **Primary**   That contained in fossil fuels in the form of coal, oil or natural gas or in nuclear energy or hydroelectricity.
- **Delivered**   That in the fuel at its point of use after allowing for extraction (or generation) and transmission losses.
- **Useful**   The portion of the delivered energy that is of benefit after allowing for the efficiency of the consuming applicance.

Gasification   See pyrolysis below.

HCFC   A hydrochlorofluorocarbon. As for CFCs above but less harmful.

L_{A10}   The level of noise exceeded 10 per cent of the time when measuring in dBA (decibels with an A-scale rating).

NR   NR (noise rating) curves are a way of using octave bands to describe a noise with a single number.

Ozone Depletion Potential   A measure of the damage caused to the ozone layer by a substance.

Pyrolysis   Gasification and pyrolysis are sometimes used interchangeably. Pyrolysis, in our case, is the breakdown of a material (typically biomass) in the absence of oxygen above 250°C. The process produces a solid (typically a char), a liquid (bio-oil) and a mixture of low-energy-content gases. Gasification is "a type of pyrolysis" and involves the transformation, through partial combustion, of a material (again, typically biomass) into a combustible gas, volatiles and an ash.

Silicon
- **P-type**   A positive P-layer of silicon.
- **N-type**   A negative N-layer of silicon.

Sound Reduction Index (SRI)   The reduction of the sound-pressure level as sound travels from one space to another through an element such as a wall, floor or roof. It is mainly dependent on the mass per unit area of the separating element. The SRI is approximately $= 20\log_{10}m + 10\,dB$ where m is the mass in kg/m^2

U-value   The rate of heat flow per unit area through an element per degree of temperature difference.

Illustration acknowledgements

The author and publishers would like to thank the following individuals and organisations for permission to reproduce material. We have made every effort to contact and acknowledge copyright holders, but if any errors have been made we would be happy to correct them at a later printing.

Cover and Preface image Photograph by Max Fordham LLP. Thanks to Llewelyn-Davies, Bellway Homes and the London Borough of Southwark

Figure 1.1   Peter Cook/View
Figure 1.2   Max Fordham LLP
Figure 1.3   The Bo01 Area Green Guide to Malmö
Figure 1.4   Max Fordham LLP
Figure 1.5   Architects' Journal
Figure 1.6   ECD Architects, "Aerial view of Cooper's Road"
Figure 1.8a   Max Fordham LLP
Figure 1.8b   Max Fordham LLP
Figure 1.9   Institute of Energy and Sustainable Development, De Montfort University

Figure 2.1   Llewelyn-Davies
Figure 2.2   Llewelyn-Davies
Figure 2.3   Llewelyn-Davies
Figure 2.4   Llewelyn-Davies
Figure 2.5   Office of the Deputy Prime Minister
Figure 2.6   Office of the Deputy Prime Minister
Figure 2.7   Greater London Authority
Figure 2.8   Greater London Authority
Figure 2.9   Greater London Authority
Figure 2.10   Greater London Authority
Figure 2.11   Greater London Authority

Figure 3.1   Courtesy Alan Baxter & Associates
Figure 3.2   Courtesy Alan Baxter & Associates
Figure 3.4   Courtesy Alan Baxter & Associates
Figure 3.5   Courtesy Alan Baxter & Associates
Figure 3.6   Courtesy Alan Baxter & Associates
Figure 3.7   Courtesy Alan Baxter & Associates
Figure 3.8   Courtesy Alan Baxter & Associates
Table 3.1   The Greater London Authority
Table 3.2   From the Urban Design Compendium, published by English Partnerships August 2002
Table 3.3   Crown Copyright (reproduced by permission)
Table 3.4   From the Urban Design Compendium, published by English Partnerships August 2002

Figure 4.1   Christina von Borcke
Figure 4.2   Christina von Borcke
Figure 4.3   Llewelyn-Davies
Figure 4.4   Llewelyn-Davies
Figure 4.5   Christina von Borcke
Figure 4.6   Christina von Borcke
Figure 4.7   Llewelyn-Davies
Figure 4.8   Llewelyn-Davies

Figure 4.9   Llewelyn-Davies
Figure 4.10   Llewelyn-Davies
Figure 4.11   Llewelyn-Davies
Figure 5.1   Peter Cook/View
Figure 5.2   David Lloyd Jones
Figure 5.3   Dennis Gilbert/View
Figure 5.4   Max Fordham LLP
Figure 5.5   Max Fordham LLP
Figure 5.6   Max Fordham LLP
Figure 5.10a   Max Fordham LLP
Figure 5.10b   ECD Architects
Figure 5.10c   Max Fordham LLP
Figure 5.12   Max Fordham LLP
Figure 5.13   Max Fordham LLP
Figure 5.14   Max Fordham LLP
Figure 5.16   Philip Vile/Haworth Tompkins Architects
Figure 5.17   Max Fordham LLP

Figure 6.1   Max Fordham LLP
Figure 6.4a & c Los Angeles County Museum of Natural History
Figure 6.4b   Marion Boyars Publishers
Figure 6.5a   University Arms Hotel, Cambridge
Figure 6.5b   Max Fordham LLP
Figure 6.6   Max Fordham LLP
Figure 6.7   Dirk Mangold, University of Stuttgart
Figure 6.8   NASA
Figure 6.9a   BEAR Architecten Gouda, The Netherlands "National Environmental Education Centre"
Figure 6.9b   Dennis Gilbert/View "Doxford Solar Office"
Figure 6.9c   Max Fordham LLP "Environmental Building" Building Research Establishment
Figure 6.10   J. Gandemer, Centre Scientifique et Technique du Batiment, A. Guyot, Groupe Ambience Bioclimatique
Figure 6.11   Dennis Gilbert/View
Figure 6.12a   Adapted from photograph and diagrams by Karle Günther/PSFU
Figure 6.12b   E&FN Spon
Figure 6.13   Mark Koehorst, Ecofys/Delft University
Figure 6.14a   "Wind measuring device" Royal Institute of British Architects
Figure 6.14b   Image by Miller Hare. With thanks also to Allies & Morrison and the Royal Institute of British Architects
Figure 6.16a   RIBA Library Photographs Collection
Figure 6.16b   RIBA Library Photographs Collection
Figure 6.16c   Janet Hall/RIBA Library Photographs Collection
Figure 6.17   Max Fordham LLP

Index

Page numbers in *italics* refer to illustrations and tables